OPENING PANDORA'S BOX

A sociological analysis of scientists' discourse

The physicist Leo Szilard once announced to his friend Hans Bethe that he was thinking of keeping a diary: 'I don't intend to publish it; I am merely going to record the facts for the information of God.' 'Don't you think God knows the facts?' Bethe asked. 'Yes', said Szilard. 'He knows the facts, but he does not know *this version of the facts*.'

Freeman Dyson, *Disturbing the Universe* (Preface)

OPENING PANDORA'S BOX

A sociological analysis of scientists' discourse

G. NIGEL GILBERT
Lecturer in Sociology, University of Surrey

and

MICHAEL MULKAY
Professor of Sociology, University of York

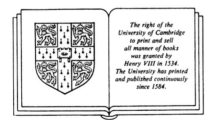

The right of the
University of Cambridge
to print and sell
all manner of books
was granted by
Henry VIII in 1534.
The University has printed
and published continuously
since 1584.

CAMBRIDGE UNIVERSITY PRESS

Cambridge
London   New York   New Rochelle
Melbourne   Sydney

CAMBRIDGE UNIVERSITY PRESS
Cambridge, New York, Melbourne, Madrid, Cape Town, Singapore, São Paulo, Delhi

Cambridge University Press
The Edinburgh Building, Cambridge CB2 8RU, UK

Published in the United States of America by Cambridge University Press, New York

www.cambridge.org
Information on this title: www.cambridge.org/9780521274302

First published 1984
Re-issued in this digitally printed version 2009

A catalogue record for this publication is available from the British Library

Library of Congress Catalogue Card Number: 83–5338

ISBN 978-0-521-25418-2 hardback
ISBN 978-0-521-27430-2 paperback

Contents

Acknowledgements

Thanks are due to authors and publishers for permission to reproduce illustrations: from J. B. Finean, R. Coleman and R. H. Michell, *Membranes and Their Cellular Functions*, Blackwell Scientific Publications, Oxford (pictures I and IX); from A. L. Lehninger, E. Carafoli and C. S. Rossi, 'Energy-linked ion movements in mitochondrial systems', *Advances in Enzymology* 29 (1967), Wiley Interscience, New York (picture II); from S. J. Edelstein, *Introductory Biochemistry*, Holden-Day, Calif. (picture III); from P. Mitchell, *Biological Reviews* 41 (1965), Cambridge University Press (picture IV); from M. D. Brand, 'The stoichiometric relationships between electron transport, proton translocation and adenosine triphosphate synthesis and hydrolysis in mitochondria', *Biochemical Society Transactions* 5 (1977), the Biochemical Society (picture V); from M. J. Selwyn and A. P. Dawson, 'Model membranes and transport systems', *Biochemical Society Transactions* 5 (1977), the Biochemical Society (picture VI); from P. Hinkle and R. McCarty, 'How cells make ATP', *Scientific American*, March 1978, 238, no. 3 (pictures VII and VIII).

Preface

The research reported in this book was conceived when a scientist friend showed us a copy of a letter written by a biochemist which seemed to indicate by its tone that there was a raging and, so we thought, sociologically interesting controversy going on in an area of biochemistry called 'oxidative phosphorylation'. Like other sociologists of science, and like our scientist friend, we assumed that part of the job of the sociologist was to strip away the formal side of science, and show what was *really* going on; an area of lively debate would, we thought, be an excellent site for such investigations.

The Social Science Research Council agreed with us, and funded a three-year research programme (HR5923) which took us around Britain and across the United States to visit and talk with the scientists who were working on oxidative phosphorylation and related topics. As we travelled, in train compartments and airport departure lounges, and especially in Howard Johnson motel rooms, we discussed together what we were hearing, and realised that although we were being given quite different accounts of what 'really was happening' by different interviewees, they *all* seemed to be plausible and, indeed, convincing. It gradually became clear that we needed to employ rather different methodological assumptions and forms of analysis from those that we had been used to in order to make any sociological sense of these data. We had to learn how to deal with variability in our accounts, in a way that recognised that the variability was not just a methodological nuisance, but was an intrinsic feature which we needed to exploit in our analyses. 'Pandora's Box' is our metaphor for the conflicting voices that spoke to us. We shall show that, nevertheless, we have been able to find order in their diversity.

Many people have helped us in our research, not least the scientists who remain anonymous in the book, but to whom we are grateful for their hospitality and kindness in talking to us. Robert Reid and Barry Gould helped us to learn the rudiments of biochemistry. Sarah Domanski did invaluable work gathering the research literature of the field. Henry Small, and the Institute for Scientific Information, provided citation data as well as entertaining us royally in Philadelphia. We must also thank Stella

Edison, Hilary Minor, Sue Plummer and Sarah Rollason for typing this manuscript and for helping with the difficult task of transcribing interview tapes of scientists talking in strange accents about their research. Jonathan Potter and Steven Yearley helped us to refine our ideas and to do justice to the intricacies of the data. Finally, we should like to thank the international community of sociologists of science, without whose arguments and objections this research would have been easier, but much less fun.

1

••

Scientists' discourse as a topic

What's in Pandora's Box?

In this book we offer a sociological analysis of material obtained from
practitioners in an area of biochemical research. If this were a typical
sociological study, we would be using that material in the following
chapters to tell the inside story about this area of social life. We would
proceed by extracting from our data what we took to be the most coherent
and comprehensive version of 'what really happened', and we would
present this story to our readers along with persuasive argument and
supporting empirical evidence.

Given that we have available a wide range of evidence about
developments within this research area, including the transcripts of
interviews, letters and other informal material, as well as access to the
formal research literature, it is likely that, as sociologists of science, we
would try to use our data to show that the area did not develop solely
through the reasoned appraisal of objective biochemical evidence; and
that a full explanation of its cognitive evolution must make reference to
the kind of social, political and personal factors documented in the less
formal sources. Having used participants' own informal talk and writings
to substantiate these claims, we would probably conclude by showing how
this case study is consistent with and contributes to a recent but steadily
growing body of sociological and historical literature on the social
production of scientific knowledge.[1]

It will be evident, however, that we do not intend to furnish that kind of
sociological analysis here. We will not be opening Pandora's Box in order
to reveal how various supposedly disreputable, non-cognitive influences
are actually at work in the field we have studied. Our reference to
Pandora's Box is not a way of referring to a supposed gap between an
orthodox view of science and the social realities revealed by sociological
research. It is, rather, a way of drawing attention to some methodological
and analytical weaknesses in previous sociological work on science.
Pandora's Box and its discordant contents are intended as a metaphor for
the remarkably diverse accounts of action and belief which appear in our

1

material and which are present, we suspect, in most sociologists' data files, but which are normally suppressed as a result of analysts' unreflective commitments to the production of a unitary 'best account' of the areas of social life they have chosen to study.[2] One of our central claims in this book is that sociologists' attempts to tell *the* story of a particular social setting or to formulate *the* way in which social life operates are fundamentally unsatisfactory. Such 'definitive versions' are unsatisfactory because they imply unjustifiably that the analyst can reconcile his version of events with all the multiple and divergent versions generated by the actors themselves.[3]

Most sociological analyses are dominated by the authorial voice of the sociologist. Participants are allowed to speak through the author's text only when they appear to endorse his story. Most sociological research reports are, in this sense, univocal. We believe that this form of presentation grossly misrepresents the participants' discourse. This is not only because different actors often tell radically different stories; but also because each actor has many different voices. In this book, we will begin to lift the lid of Pandora's Box in order to give some of these voices the opportunity of being heard.

In the rest of this chapter, we will develop this argument with respect to science, to show that the goal of constructing definitive analysts' accounts of scientists' actions and beliefs is possibly unattainable in principle, and certainly unattainable in practice as long as we have no systematic understanding of the social production of scientists' discourse. Sociologists, historians and philosophers have been able to document and make plausible so many divergent analyses of science (and continually undermine each other's claims) because scientists, the active creators of analysts' evidence, themselves engage in so many kinds of discourse. Thus we recommend that analysts should no longer seek to force scientists' diverse discourse into one 'authoritative' account of their own. Instead of assuming that there is only one truly accurate version of participants' action and belief which can, sooner or later, be pieced together, analysts need to become more sensitive to interpretative variability among participants and to seek to understand why so many different versions of events can be produced.

We will try to show that, analytically, there is much to be gained by opening Pandora's Box in the sense of setting free the multitude of divergent and conflicting voices with which scientists speak. Of course, the interpretative variability found in scientists' discourse undoubtedly occurs in other areas of social life.[4] Consequently, our attempt in this book to reorient the sociological analysis of science in order to cope with the variability of participants' discourse has obvious parallels with and implications for other fields of sociological inquiry. Although it would be distracting if we were continually to draw attention to similarities between the sociological analysis of science and that of other areas of social life, it is important to stress that in this book we begin with analytical and methodological concerns that are in no way peculiar to the sociology of science and that our conclusions about the importance of discourse analysis should apply to any realm of sociological study.

Analysis of social action in science

In this section, we demonstrate why there is a need for a form of sociological analysis which focuses on the organisation of scientists' discourse. Let us start by looking at a study of scientists undertaken about ten years ago by Marlan Blissett.[5] We have chosen to comment on Blissett's analysis because it clearly shows how sociological interpretation of social action typically depends heavily on unexplicated interpretative work carried out by participants and embodied in their discourse. Another reason for choosing this study is that the research network examined by Blissett overlaps considerably with the one with which we are concerned in this book. Differences between Blissett's study and our own are therefore unlikely to be due to major differences in kinds of respondent or kinds of data. They are more probably signs of genuine differences in analytical approach. A brief description of his work will therefore help us to clarify the distinctive features of our approach to analysis.

Blissett focuses on the role of politics in science. His main thesis is that it is a myth that scientists are neutral and disinterested actors when they engage in research. He aims to show that the professional actions of scientists are essentially political in character and that scientists regularly engage in such political manoeuvres as 'marketing, salesmanship and manipulation'. He suggests that these activities are not regrettable and infrequent lapses by otherwise disinterested scientists, but are vital aspects of the process of scientific enquiry.

Such a thesis is not only sociologically interesting, but it is also typical of much sociological analysis in formulating definitive categorisations of participants' actions. Thus Blissett proposes that some actions were political, as distinct from any other type of action. Blissett further claims that such political actions direct scientific perception and influence the acceptance or rejection of specific theories and ideas. Here Blissett also typifies much sociological analysis in suggesting that his categorisations identify stable entities which cause other social phenomena. Furthermore, Blissett asserts that the material from his interviews with biochemists clearly substantiates these claims.

Blissett's conclusions describe scientists' actions and the social consequences of these actions. However, Blissett's data consist entirely of statements obtained from interviews with scientists or from their written descriptions of the field. In other words, his data are *accounts* of action. We need to examine how Blissett manages to derive conclusions about actions and their consequences from participants' accounts. We can best do this by quoting some examples of his data and analysis.

Blissett begins by noting that research in the area he is studying is pervaded by controversy and that 'the importance of controversies of this nature is that they are unlikely to be resolved by appeal to evidence alone'. This observation is supported by a quotation from an article by one of the contenders in the controversy, who writes that:

> Until a few years ago the conceptual framework in the field [under study] played a relatively minor role in determining the direction and in shaping the design of experimentation. In the phase of describing phenomena, the conceptual framework is not crucial. It is only when experimentation reaches the interpretative and exploratory stages that *permissiveness* or *indifference* with respect to the conceptual framework has consequences which inhibit progress. [Emphasis added][6]

Thus Blissett proposes that the resolution of this controversy depends on more than the scientific evidence and he justifies this claim by quoting a scientist who, in talking of the role of permissiveness and indifference, can be seen to be saying much the same thing.

Blissett organises his analysis to show that a crucial additional factor which helps to determine the outcome of controversy is the effect of political strategies. It was this concern with political action, he states, which led him to select this particular field as one in which interviews might yield valuable data. He writes, 'The prospect of a hard-bitten scientific controversy led to interviews concerning the political nature of the matter with biologists in the field.'[7] During these interviews, Blissett was offered many statements which provide prima-facie evidence in support of his notion that political strategies are important. For example, one of the contenders in the dispute, he says, admitted that the present level of theoretical conflict in biology was unequivocally immersed in the political strategies of personal salesmanship and scientific advertisement. The same scientist is quoted as stating that: 'To make changes you have to be highly articulate, persuasive, and devastating. You have to go to the heart of the matter. But in doing this you lay yourself open to attack. I've been called fanatical, paranoid, obsessed . . . but I'm going to win. Time is on my side.'[8]

Further evidence of a similar nature is provided. For instance, Blissett presents another lengthy quotation in which the speaker says, in part:

> However, aside from the technical difficulties that attend his theory, [he]
> himself must be held responsible for some of his 'selling' problems.
> Labelled by a colleague of mine as 'insulting', [he] indeed hardly
> possesses the patience necessary for the presentation of his theories . . .
> He suffers not so much from a repressive oligarchy bent on his
> destruction, as from a plurality of opponents, some of whom like the idea
> of a revised membrane model, but who detest the man who is responsible
> for its initial formulation.[9]

On the basis of a series of such quotations, in which participants characterise their own and/or others' actions in 'political' terms, Blissett claims to have shown that political action occurs frequently in this area and that it has significant consequences.

We have presented sufficient material in this brief résumé of Blissett's study to be able to draw out the basic elements of the interpretative method he uses. His procedure is to make a claim about scientific action, such as that it is political, and then to confirm this claim by presenting material in which scientists themselves can be seen to be making the same claim. Thus to justify his thesis that political action is involved in the creation of scientific knowledge, Blissett offers passages in which scientists describe their own and others' actions as political. In practice, such passages are not difficult to find; they occur regularly in our own interview transcripts, for example. Then, having shown that descriptions of political activity do often appear in statements made by scientists, Blissett concludes that political action is a fundamental feature of science.

We can set out this form of analysis in a more systematic way as a series of steps. This is worth doing because not only Blissett's but most qualitative studies seem to follow these steps.

(1) Obtain statements by interview or by listening to or observing participants in a natural setting.

(2) Look for broad similarities between the statements.

(3) If there are similarities which occur frequently, take these statements at face value, that is, as accurate accounts of what is really going on.

(4) Construct a generalised version of these participants' accounts of what is going on, and present this as one's own analytical conclusion.

This is not an unreasonable characterisation of Blissett's procedure. He interviewed scientists (1), found numerous statements which dealt with salesmanship, manipulation and the like (2), took it that these statements were accurate reports of the way in which scientists acted (3) and concluded that science is political (4).

Blissett's use of participants' accounts is far from unique. As has been shown elsewhere, procedures very similar to that adopted by Blissett are recommended in influential pronouncements on sociological methods and employed in qualitative studies of quite different types of social actors;[10] and a wide range of empirical studies specifically within the sociology of science, quantitative as well as qualitative, have been shown to be as heavily dependent as that of Blissett on interpretative work carried out by participants.[11]

This does not mean, of course, that the analyst does *nothing but* reproduce participants' discourse. Analysts do typically make contributions of at least three kinds. They subsume participants' specific pronouncements under more general concepts. Blissett does this, for example, when he collects together a variety of particular statements referring to manipulation, influence, manoeuvring, and so on, as all about one kind of action, namely political action. At the same time, analysts tend to generalise participants' statements about particular actors or actions to whole classes of social action and to whole groups of actors. Thirdly, analysts identify those segments of participants' discourse which are to be regarded as accurately representing important social processes occurring within the area of social life under study. Other parts of participants' discourse are ignored or treated as inaccurate. Although these three facets of sociological research practice are closely related, we will concentrate on the third component. In the sections which follow, we will show that there are good theoretical as well as practical reasons for doubting whether some sections of participants' discourse can be selected as providing sociologically more satisfactory descriptions of members' action or belief than others.

The context-dependence of participants' discourse

The difficulty with taking any collection of similar statements produced by participants as literally descriptive of social action is the potential variability of participants' statements about any given action. The reasons why we would expect participants' statements to be potentially variable are clearly expressed by Halliday in his discussion of the basic characteristics of language use.

> The ability to control the varieties of one's language that are appropriate to different uses is one of the cornerstones of linguistic success ... Essentially what this implies is that language comes to life only when functioning in some environment. We do not experience language in isolation ... but always in relation to a scenario, some background of persons and actions and events from which the things which are said derive their meaning ... any account of language which fails to build in

the situation as an essential ingredient is likely to be artificial and unrewarding . . . *All* language functions in contexts of situation, and is relatable to those contexts. The question is not what peculiarities of vocabulary, or grammar or pronunciation, can be directly accounted for by reference to the situation. It is *which* kinds of situational factor determine *which* kinds of selection in the linguistic system . . .[12]

We do not wish to endorse in detail every aspect of Halliday's treatment of the relationship between linguistic variation and social context. Nevertheless, we take his general claim with respect to the complex interdependence between participants' discourse and its situation of production to be firmly established. If there *is* a strong connection between the form and substance of discourse, on the one hand, and the social situation in which discourse is produced, on the other hand, it follows that discourse can never be taken as simply descriptive of the social action to which it ostensibly refers, no matter how uniform particular segments of that discourse appear to be. For similarities between different statements are just as likely to be the consequence of some similarity in the context of linguistic production as of similarity in the actions described by those statements. For instance, the apparently overwhelming orientation towards political action in Blissett's material may well have been at least partly due to a response by interviewees to unintentional cues provided by the investigator. Without detailed examination of the linguistic exchanges between researcher and participant, and without some kind of informed understanding of the social generation of participants' accounts of action, it is not possible to use these accounts to provide sociologically valuable information about the actions in which analysts like Blissett are interested. It certainly cannot be assumed that marked similarities within such collections of statements indicate the existence of corresponding regularities in social action.

Traditional sociological research, like that exemplified in the previous section, operates according to a methodological principle of linguistic consistency; that is, if a 'sufficient proportion' of participants' accounts appear consistently to tell the same sort of story about a particular aspect of social action, then these accounts are treated as being literally descriptive.[13] Only in those instances where the existence of incompatible accounts is treated as sociologically significant do analysts pay attention to the social generation of accounts; and in such cases, reference to the social or personal context of participants' discourse is usually introduced into the analysis in order to explain away those accounts which weaken the analyst's conclusions, on the grounds that they are exaggerations, biased reports, ideology, lies, and so on.[14] Acceptance of Halliday's argument, however, implies a need to revise such an approach to participants' discourse in a fundamental way. For Halliday proposes that there are no literal descriptions available and that all linguistic formulations, indeed all members' symbolic products,[15] have to be understood in relation to their context of production. This proposition clearly implies that the systematic investigation of participants' discourse is methodologically prior to analysts' *use* of such discourse to characterise and explain social action. Even more significantly, it may be that the traditional sociological goal of providing analyses of social life which build upon the interpretations furnished by participants is made unattainable by participants' ability to engage in the creative use of language.

Direct observation and participants' discourse

Proponents of traditional methodologies might respond to the argument so far in one of two ways. In the first place, they might accept that participants' retrospective accounts of action and belief, as obtained for example from interviews, autobiographies, review articles, public lectures, and so on, are highly variable, context-dependent, and therefore unreliable; but they might suggest that it is possible to replace such indirect sources of data with direct observation of social action as it occurs. Some of the recent ethnographies of work in scientific laboratories seem to exemplify this view.[16] The idea is that by observing actions as they take place, the analyst is able to avoid, or at least reduce to an acceptable minimum, any dependence on participants' potentially variable interpretative activities.

Although we have no wish to deny the interest of this kind of observational work, it does not seem in itself to resolve the difficulties identified above. There are several reasons why this is so. First, social action is not 'directly observable'. The observable, physical acts involved in performing an experiment, for example, do not reveal whether the experiment is an attempt to refute an hypothesis, an attempt to find a new way of measuring a known variable, a routine check on the experimental apparatus, and so on. Which of these or other actions is being observed on any particular occasion can only be established by reference to the statements, either written or spoken, of participants. Yet, not only can descriptions of an experiment vary considerably from one scientist to another,[17] but the accounts given of a particular experiment by an individual scientist can, as Hanson and others have shown, vary appreciably.[18] Thus so-called 'direct observation' of social action as it takes place in no way frees the observer from reliance on the potentially variable discourse of participants.

The ability of social actors to characterise a given set of activities in various different, and sometimes apparently incompatible, ways becomes understandable if we accept that social activities are the repositories of multiple meanings. For instance, does a given set of activities constitute an experiment, an attempt indirectly to raise more research funds, an effort to secure professional credibility, a bid for more students; or can it be any or all of these, depending on the context in which the actor is talking or writing about his actions? If the latter is the case, and we suggest that it is, then 'the meaning' of his action is variable and context-dependent. It will be quite impossible to establish the nature of the action unequivocally by being present at and directly observing the original laboratory experiment. For the social character of the original laboratory work will continually change as participants interact in different settings and thereby generate different kinds of linguistic gloss upon those initial activities.

It seems best, then, to conceive of the meaning of social action, not as a unitary characteristic of acts which can be observed as they occur, but as a diverse potentiality of acts which can be realised in different ways through participants' production of different interpretations in different social contexts. It is important to recognise that this production of social meanings through language is a temporal process. Actors continually reinterpret given actions as their biography unfolds and as changing circumstances lead them to fit these actions into new social configurations. And the meaning of each new situation is defined in part through participants' reinterpretations of what they have done in the past.[19] Consequently, participants' observable accomplishment of actions at a specific point in time cannot be neatly distinguished from, or separated from, the kind of retrospective story-telling which is generated in interviews and other indirect methods of data collection. The technique of direct observation cannot avoid becoming entangled in members' variable and context-dependent reconstructions of their social world, because this kind of reconstruction is a pervasive feature of the creation of social meaning.

These reflections on the nature of direct observation thus serve only to strengthen our previous argument for the methodological priority of analysis of participants' discourse. However, exponents of traditional methodologies might still reject the argument we are developing on the grounds that, even though all participants' statements are socially generated, this does not mean that some statements by participants are not more accurate or more sociologically useful than others. For instance, it has been argued that a scientist's rendering of 'The Bluebells of Scotland' or a page torn at random from a telephone directory are obviously less informative about the nature of social action in a research network than a page of detailed interview transcript or copies of letters exchanged among participants.[20]. It is proposed in this line of argument that sociologists can tell good from bad accounts of action and belief; and that they do so by acquiring tacit craft skills which enable them to assess the veracity of different kinds of account.[21]

This view of social research is obviously unsatisfactory if one has reached the conclusion, suggested above, that the social world is not composed of a series of discrete, one-dimensional actions which can be more or less accurately represented. Once we begin to conceive of the social world in terms of an indefinite series of linguistic potentialities which can be realised in a wide variety of different ways and which are continually reformulated in the course of an ongoing interpretative process, the simple procedure of sifting good from bad accounts becomes entirely inappropriate. But even if we remain within the traditional conception of social action, this line of argument still has several weaknesses. For example, the fact that all researchers distinguish fairly easily between relevant and irrelevant data, between participants' letters and the telephone directory, in no way implies that the analysis of relevant data can be accomplished with equal facility. Moreover, it is clearly being conceded that sociological interpretation does depend on the analyst's capacity for understanding and systematically allowing for the social generation of participants' discourse. What is being rejected is the idea that this topic could or should become a critical focus of sociological investigation. There seems to us to be no good reason to insist that such a crucial facet of the sociological craft could not be considerably improved by means of careful, explicit study. In addition, we suggest that linguistic variability is much greater than is implied in the view summarised above; so much so, that no degree of craftsman's expertise can enable the sociologist to sort out the interpretative dross within participants' discourse from what is sociologically valuable.

The variability of participants' discourse

This last claim clearly requires, and is open to, empirical demonstration. If participants' accounts of action and belief are so variable that, when this variability is acknowledged and systematically considered, it prevents the construction of satisfactory sociological interpretations, then it should be possible to demonstrate this by reference to empirical data. However, we can hardly formulate a convincing case for such a general argument in the present introductory chapter, before we have even begun to provide the background information necessary for an understanding of our data. Furthermore, it will not be particularly helpful to examine one or two brief illustrations of actors' interpretative variability at this juncture, for they could easily be dismissed as instances of unusually awkward data selected to make our point. We have tried to deal with this difficulty in a series of papers published elsewhere. In these papers we have worked systematically through a batch of material on scientists' theory-choice.[22]

The aim of these papers was to use multiple samples of data on a single topic from the same collection of material in order to display in detail just how diverse are participants' responses in relation to a narrow range of social action. Considerable space is required to achieve this task and we will attempt nothing along these lines in this chapter. However, the chapters which follow will, among other things, also provide empirical confirmation of the point. Furthermore, other authors are beginning to recognise the importance of the variability of actors' discourse, to furnish evidence that it occurs in many realms of social life and to attempt to deal with it analytically.[23] In this introduction, therefore, we will limit ourselves to bringing out some of the implications of interpretative variability through continuing the comparison of our analysis with Blissett's.

As mentioned above, there is every reason to expect that Blissett's data should be very similar to our own. However, whereas Blissett accepts participants' characterisations of their own or others' actions at face value, we find that, when we look at any collection of participants' characterisation on a given topic from our data, almost every single account is rendered doubtful by its apparent inconsistency with other, equally plausible, versions of events. The degree of variability in scientists' accounts of ostensibly the same actions and beliefs is, in fact, quite remarkable. Not only do different scientists' accounts differ; not only do each scientist's accounts vary between letters, lab notes, interviews, conference proceedings, research papers, etc.; but scientists furnish quite different versions of events within a single recorded interview transcript or a single session of a taped conference discussion.

We are not suggesting that participants' varying accounts are 'intrinsically incompatible'. It is presumably always possible for the analyst, like the participant, to extract a 'definitive' version of events from even the most diverse set of accounts: for example, by restating what particular respondents 'really meant' in the light of their statements elsewhere, by eliminating certain statements as hyperbole, irony, rhetoric, etc., or by interpreting the data in accordance with tacit understandings gleaned in the course of interaction with participants. But our experience is that this process of reinterpretation to distil a comprehensive, ultimate version can produce firm conclusions only by disregarding copious interpretative uncertainties.

Consider a hypothetical example. Scientist A states, on one occasion, that certain of B's actions were, in Blissett's sense, political. Scientist C, without being questioned specifically on this point, portrays B's actions quite differently, as being dominated by a selfless pursuit of scientific truth. Whereas Scientist B, who naturally provides more detail about his own actions than the other speakers, appears to be somewhat inconsistent, giving some accounts which seem to support C's testimony, but also furnishing evidence which favours A.

One possibility is for the analyst to try to check things further by asking these respondents to reconsider what they have said. However, this is just as likely to make things worse as better. The analyst may well find that Scientist A now has second thoughts and denies that he really knows much about the nature of B's actions. As a result, the analyst is forced to reconsider whether he can use *any* of the material provided by A. For this respondent, who previously appeared to be an entirely reliable witness, now seems untrustworthy. Thus the analysis of any other topics on which A was a crucial informant has been put in jeopardy.[24]

A second possibility is simply to discount some of the available testimony, in order to obtain a consistent residue. But this is difficult because, whichever choice we make, we will have to reject part of B's evidence, whilst accepting some other part of it. As with A, we are now faced with the danger of accepting evidence from a basically unreliable witness. Moreover, we cannot separate the acceptable from the un-acceptable accounts of B's actions without also treating as unproblematic our analysis of another batch of data, though it is likely to be just as variable as that dealing with B's political actions. For instance, we could only discount A's testimony as being distorted, say, by his intellectual rivalry with B, if we were able to establish unequivocally that such rivalry existed and that it influenced A's accounts, but not those proffered by B that we have accepted.

We will not prolong our discussion of this hypothetical example. We offer it simply as a condensed illustration of the practical difficulties arising in the course of attempts to carry out traditional forms of analysis of complex qualitative data. Although this example is obviously constructed in a way which emphasises the problems of reconciling participants' accounts of action, it does not exaggerate these problems. Indeed, the difficulties that can be illustrated through an example employing three respondents are but a pale reflection of those which occur when thirty or more scientists are involved. We suggest that each attempt by the analyst at reconciliation, if it is checked carefully against other material and against the analyst's other interpretations, regularly provokes further interpretative problems. Furthermore, as we have indicated several times before, any attempt by the analyst to escape from this potentially endless sequence of interpretative revisions, involves him in relying on unsubstantiated assumptions about the social generation of actors' discourse. Once again it becomes clear that the traditional analysis of social action cannot be successful without a systematic understanding of the production of discourse.

Interpretative uncertainties of this kind, which constitute major problems for analysts, do not pose any great difficulty for participants, who have at their disposal a range of flexible techniques which enable them to make sense of whatever is going on in a way that is adequate for most practical purposes. But this kind of everyday reasoning is not sufficiently grounded in data to be satisfactory for analytical purposes. Moreoever, gathering more data does not help. The more data one has, the more intractable is the task of analytical reconciliation.

Some readers may wonder why, if scientists' interpretative practice is so variable, it has not been regarded as a serious difficulty until now. One reason is that detailed examination of scientists' accounts is quite recent. Secondly, as we noted above, it is always possible to extract plausible versions of events from qualitative data, so long as the analyst's interpretative practices are not subjected to detailed scrutiny. This is what has happened in the past. Moreover, analysts' versions are typically illustrated by appropriate selections from participants' accounts, without the reader having access to any of the alternative versions which are produced by all social actors in great numbers but which are either unrecorded or ignored or explained away by analysts committed to producing definitive versions of their own.

For traditional sociological analysis of social action, then, participants' interpretative variability causes fundamental, and perhaps insoluble, difficulties. In this book we intend to begin to develop an alternative form of analysis which turns this intractable methodological liability into a productive analytical resource. We refer to this form of analysis as discourse analysis.

Discourse analysis

The central feature distinguishing discourse analysis from previous approaches to the sociology of science is that, in the now familiar phrase, it treats participants' discourse as a topic instead of a resource. Previous approaches have been designed to use scientists' symbolic products as resources which can be assembled in various ways to tell analysts' stories about 'the way that science is'. Discourse analysts, in contrast, begin from the assumption that participants' discourse is too variable and too dependent on the context of its production to be amenable to this kind of treatment. At least initially, they abandon the goal of using scientists' discourse to reveal what science is really like, and concern themselves instead with describing the interpretative methods which are used, not only by participants but also by traditional analysts, to depict scientific action and belief in various different ways. Instead of taking as the initial question, 'How can a definitive analysts' version of action and belief be extracted from scientists' variable discourse?', discourse analysts concentrate on what appears to be the methodologically prior question, 'How are scientists' accounts of action and belief socially generated?'[25]

Discourse analysis, then, unlike the kind of analysis exemplified above by means of Blissett's study, does not seek to go beyond scientists' accounts in order to describe and explain actions and beliefs as such. It focuses rather on describing how scientists' accounts are organised to portray their actions and beliefs in contextually appropriate ways. Thus, discourse analysis does not answer traditional questions about the nature of scientific action and belief. What it may be able to do instead is to provide closely documented descriptions of the recurrent interpretative practices employed by scientists and embodied in their discourse; and show how these interpretative procedures vary in accordance with variations in social context. Discourse analysis, then, is the attempt to identify and describe regularities in the methods used by participants as they construct the discourse through which they establish the character of their actions and beliefs in the course of interaction.

This change in analytical focus has several significant implications for sociological practice. In the first place, it means that analysts can stay much closer to their data. The traditional concern with social action often required the analyst to infer the nature of past actions from participants' statements about those actions. Discourse analysis, in contrast, assumes that such statements are versions of events which are to be understood in relation to the context in which they are produced. In this sense, scientists' verbalisations are no longer used as indirect indicators of something else which is held to be more sociologically interesting. Scientists' discourse, its organisation and contextual production, become the object of sociological investigation. Secondly, the new approach makes it clear that no particular class of participants' discourse is to be taken as analytically prior. The informal talk whereby actions and beliefs are constituted at the laboratory bench is not regarded as having primacy over any subsequent reinterpretations around a coffee table, at a conference, in a research paper or in an interview. Thus the analyst is in principle able to allow for the variability of scientists' discourse and to seek to understand it in relation to variations in social context. One potential advantage of this approach, therefore, is that it should help us to appreciate how the various analytical conclusions to be found in the sociological literature have arisen from analysts' use of different kinds of scientific discourse for their data.[26]

Thirdly, analysis of discourse frees the analyst from direct dependence on participants' interpretative work. The task of the analyst is no longer to reconstruct what actually happened from scientists' attempts to portray their own and their colleagues' actions and beliefs, but to observe and reflect upon the patterned character of participants' portrayals. The latter is only occasionally a topic of interest to scientists themselves. By distinguishing in this way between analysts' and participants' objectives, the latter's accounts become more clearly available as a topic rather than as an unexamined analytical resource.

Fourthly, whether or not discourse analysis is necessarily a replacement for traditional analysis, it does seem clear that it is a necessary prelude to the satisfactory resolution of traditional questions. Given that participants' use of language can never be taken as literally descriptive, it seems methodologically essential that we pay more attention than we have in the past to the systematic ways in which our subjects fashion their discourse. Traditional questions, it seems to us, will continue to remain unanswered, and unanswerable, until we improve our understanding of how social actors construct the data which constitute the raw material for our own interpretative efforts.

A fuller exposition of what we mean by discourse analysis will unfold in the following chapters. But in order to avoid unnecessary confusion at the outset, let us briefly compare our form of analysis to certain somewhat similar enterprises which are often given the same name. The point that we wish to make is that our analytical project is supplementary to these other branches of discourse analysis.

Our work supplements most prior work on the social organisation of discourse in being directly concerned from its inception with science and scientists. For very little systematic study of scientific discourse has so far been undertaken; although there are clear signs that a rapid growth in this kind of research is under way.[27] The chapters which follow, therefore, can be seen as an attempt to open up for systematic investigation an area of social and technical discourse which is of major significance within the culture of present-day society.

The second way in which our analysis supplements prior work on discourse can be clarified by comparing it briefly with two very different studies, which represent the two ends of a spectrum of discourse analysis. In their book *Official Discourse: On Discourse Analysis, Government Publications, Ideology and the State*, Burton and Carlen[28] take over some ideas from the continental tradition of discourse analysis, and particularly from the writings of Foucault, in order to show that the language employed in official publications is a language of class domination operating as an ideological legitimation of the state. This analysis, although it reproduces material from government reports and uses such material to illustrate its conclusions, does not examine passages of discourse in detail. The analysis is focused, not on small-scale linguistic variation in official texts, but upon those textual components which can be plausibly linked to politically significant features of the structure of contemporary society.

In the chapters below, we will not attempt to emulate Burton and Carlen's ambitious form of analysis. We will not try to explain the nature of scientific discourse by presenting it as an outcome of the actions of dominant social groups. Nor will we try to establish connections between scientific discourse and the wider structure of society. These analytical objectives resemble those characteristic of most traditional sociological research and are unacceptable to us for much the same reasons. Another difference between our approach and that of Burton and Carlen is that we will not employ the abstract terminology of the speculative discourse theorists. Instead of applying an abstract, preconceived language to our data in order to show how discourse arises from and reproduces complex social structures, we will begin with an examination of those terms and interpretative features which seem to arise naturally in the course of participants' own discourse and we will extend our analysis to cope with collective or structural phenomena only to a limited extent, that is, only in so far as it seems to us to be possible to provide detailed evidence for the analytical claims being advanced.

Despite these differences between our approach and that of Burton and Carlen, their work is similar to ours in its treatment of discourse as an important facet of social life and in its attempts to specify some of the features of discourse which recur in particular kinds of social context. However, our work also resembles in several respects a very different approach to discourse analysis, that of British sociolinguistics. This tradition is well exemplified in the collection of papers, *Studies in Discourse Analysis*, edited by Coulthard and Montgomery.[29] This work is similar to ours in that it attempts to provide a systematic description of the discourse employed by particular groups of social actors in specific settings. The analysis is very closely based on data and every attempt is made to cope with subtle variations in linguistic usage. Although this corpus of research began with a study of interaction and turn-taking in classrooms, it has been elaborated to cope with a range of other social situations and moves have been made towards analysing the structure of extended monologues, such as lectures in science and engineering, and of formal texts, such as science textbooks.[30]

Once again, however, despite certain similarities in approach and despite the tendency of the sociolinguists to become increasingly interested in science, their work is significantly different from our own. For example, we have been concerned from the outset, not with short conversational exchanges, but with complex, extended passages of discourse such as responses in informal interviews, letters and research papers. Accordingly, unlike the sociolinguists, we have paid less attention to the nature of interaction and transitions between speakers, and considerably more attention to devising a form of analysis which can make sense of the content of comparatively lengthy stretches of uninterrupted discourse. One result of this has been that we have not sought to develop a systematically articulated set of concepts which meshes neatly with the terms of grammatical theory. In our view, our data contain too much that is as yet unanalysed to make any move towards conceptual closure appear worthwhile. Another significant difference between ourselves and the sociolinguists is that we use our analysis of discourse to try to provide new insights into what sociologists would call the collective phenomena of science.[31] But the phenomena which we examine, such as consensus and humour, are on a lower level of abstraction than those investigated by Burton and Carlen. On the other hand, we undertake no analysis which is as fine-grained as, for instance, that of Brazil on variations of intonation.[32]

Thus, in various ways, the analysis below occupies a middle ground within the domain of discourse analysis. It represents part of an eclectic movement toward the systematic investigation of discourse in all areas of social life. As a contribution to that movement it serves as a possible bridge between the sociolinguistic and the sociostructural approaches to discourse. At the same time, this study is an attempt to explore an alternative to traditional sociological methods of research on social action and belief. It is also intended to begin to show how an investigation of discourse can cope with the analysis of collective phenomena. Finally, it is a study of aspects of scientific culture which documents some of the methods used by scientists as they continually construct and reconstruct their social world.

2

•••

A possible history of the field

Although in this book we aim to obtain general conclusions about scientific discourse, most of the data we shall be drawing upon come from the interviews that we carried out with scientists working on bioenergetics, and from their research papers and other literary products. This chapter outlines the recent history of bioenergetic research, and at the same time provides an overview of the scientific issues involved. These should help in understanding the material with which we shall be illustrating our later discussion.

Biochemists study the chemistry of living matter; bioenergetics, a research specialty of biochemistry, deals with the organic processes that create, transport and store chemical and other kinds of energy. The scientists with whom we shall be concerned are particularly interested in the formation of a complex molecule called ATP (adenosine *tri*phosphate), which plants, animals and bacteria use as a means of moving and temporarily storing energy within the cell. The process whereby this molecule is formed in animals and bacteria is called oxidative phosphorylation (often abbreviated in informal speech to 'ox phos'); in plants, the process is somewhat different and is known as photo-phosphorylation.

ATP is created within the cell from the combination of ADP (adenosine *di*phosphate) and inorganic phosphate in conjunction with an enzyme, ATPase, a process which requires an input of energy. The reaction may alternatively proceed in the reverse direction, yielding ADP and inorganic phosphate from ATP, with the production of energy that can be used for other cellular processes. Since the late 1940s, it has been known that the creation of ATP in animals takes place in small particles called mitochondria. These are located within the cell protoplasm. The mitochondria appear as bodies enclosed by two membranes, the inner one of which is highly convoluted. It is now widely accepted that it is the inner of the two membranes that is essential for the formation of ATP by oxidative phosphorylation.

During the late 1940s and early 1950s, evidence accumulated to show that a series of chemical oxidation and reduction reactions linked together components of the inner mitochondrial membrane into a chain. The operation of this 'respiratory chain' (often called the 'electron chain' by our respondents) seemed to be coupled to the formation of ATP. Nevertheless, the details of the components of the chain, the reactions which took place in it, and the precise relationship between these reactions and the formation of ATP were not then clear. However, the biochemical details had previously been established with reasonable confidence for another ATP-forming reaction, 'substrate level phosphorylation', and it seemed that analogies might reasonably be drawn between this well-understood reaction and oxidative phosphorylation.

This chapter will mainly be devoted to relating how bioenergeticists, over the course of three decades, came to understand in finer detail the mechanism of oxidative phosphorylation. The material we shall be drawing on to tell this history consists principally of the interviews we conducted in 1979 and 1980 with 34 bioenergeticists. This sample constituted approximately 50 per cent of those British and American scientists who have published more than one or a few occasional papers in the area. The scientists were interviewed, usually in their laboratories, for on average between 2½ and 3 hours. The interviews were tape-recorded and transcribed in full. We then read through the transcripts and copied those pages which included material relating to the topics which interested us. The passages from the interviews concerning each topic were placed together in 'topic files', so that we had convenient access to all the material on, for instance, consensus or diagrams and pictorial representations.[1] We aimed to make each file as inclusive as possible so that no passage which could be read as dealing with a particular topic was omitted from its file.

Other materials we drew on for the work reported in this book include a collection of letters circulated privately amongst the major figures in the field, a listing of research papers obtained from the Science Citation Index of those co-citation clusters[2] which we identified as relating most closely to bioenergetics, a collection of some 400 articles from the bioenergetics research literature (including all the review articles published between 1960 and 1980), copies of the relevant portions of around 30 biochemistry textbooks, and curricula vitae obtained from our interviewees.

In this chapter, we intend to present only an outline of the history of bioenergetics. Working at this high level of generality, it is relatively easy to piece together, from what the participants told us, an apparently coherent and plausible history of scientific developments. But, as we shall note at intervals, this coherence is a fragile construction, obtained at the expense of ignoring the variations and inconsistencies in the accounts we were offered. As we have suggested in the first chapter, and will go on to demonstrate in some detail in later chapters, there is a very considerable degree of variability in the accounts we obtained. For the purposes of this introductory history, we have chosen to ignore or suppress most of this variability. In doing so, we have probably followed the procedures used in most other studies of science, and by the participants themselves, in resolving potential conflicts of evidence by appeal to common-sense notions of what was most likely, or most easily understandable. We have adopted this strategy for this chapter alone because our main intention here is not to provide an analysis of the scientists' discourse, as it is in the rest of the book, but to provide the necessary background for an appreciation of the talk of our bioenergeticist respondents. Towards the end of the chapter we shall make some brief comments about the kinds of variability in the accounts we have drawn on, though this will be a topic we shall explore in much greater detail in later chapters.

Before beginning to relate this history, we must deal with two preliminary issues of nomenclature. We have reluctantly adopted the convention of referring throughout to scientists as 'he', partly because the phrase 'he or she' becomes clumsy on repetition, but also because all the scientists we interviewed for the study were male. We have also adopted the sociological convention of using pseudonyms to refer to our respondents. In general, we have applied these pseudonyms consistently throughout the text, so that the scientist called Barton in this chapter will also be called Barton elsewhere in the book. There are, however, some exceptions. Wherever it has seemed to us at all possible that a participant's real name could be discovered by a persistent detective – for example, where the author of a traceable research paper is mentioned in the text or where speakers' names appear in a diagram which is well known to participants and which is publicly available, then new pseudonyms have been introduced for the relevant passages.

In one case, however, it has proved to be impossible to maintain a participant's anonymity. The dominant theory in this area of research, the chemiosmotic theory, is very closely associated with one particular scientist whom we have called Spencer. Because this scientist is so central to the field, because he is mentioned so frequently in our material and because his name is used eponymously for the chemiosmotic theory, it would have been immensely confusing if we had tried to vary his name in order to cover his tracks. We have, therefore, used the pseudonym of Spencer consistently throughout the book. As a result, there can be no doubt that those already familiar with the field will know who Spencer is. But this is not a mere pretence of anonymity, for most of our readers will not be biochemists, they will not be familiar with the field, and, we suspect, few will want to know Spencer's real name. Thus, there is nothing to be gained by abandoning the convention of using a false name in the case of Spencer. Moreover, the systematic use of pseudonyms in the text helps to emphasise in a formal manner our contention that the passages of discourse we examine in subsequent chapters are not of interest as statements by Barton or Spencer or any other specific scientist, but as instances of generic interpretative procedures which are regularly used by scientists.

A version of the history of research on oxidative phosphorylation

Although forebears of the research area are traced by some back to the 1920s and 30s, most commentators see discoveries made in the 1940s as first opening up the field. One respondent, for example, began his account of the origins of work on oxidative phosphorylation by referring to research on pyrophosphates, a generic name for compounds like ADP and ATP.

> **2 A**
> Early in the 40s, a Russian and Lippmann in America had shown there was a general form in which energy was used and that was a molecule called pyrophosphate. It's called ATP, adenosine triphosphate, and triphosphate means just that phosphates are linked together in a chain . . . If you put two phosphates together, those two phosphates have to be forced together by energy, so the energy currency running the system was going to be this pyrophosphate . . . [Jennings, 13]

Also in the 1940s, Keilin in Cambridge had isolated the respiratory chain and some of its components, and American biochemists had discovered the mitochondrion within animal cells' protoplasm. Grant, then a young scientist who had been involved in the discovery of the mitochondrion, had had the opportunity to establish a new research laboratory:

> **2 B**
> I decided that this was my life's work – the mitochondrion. Here is the system that made ATP and fatty acid oxidation. You name it, the mitochondrion did it! And this was a worthy subject for a lifetime's effort. So we were very fortunate. Rockefeller gave us lots of money. We set up a beautiful laboratory – everything you want, large-scale isolation facilities. It turned out mitochondria by the bucket. And then there was a golden period of twenty years, where we just did all the basic work, isolated it, tore it apart, separated out the complexes, studied the citric acid cycle, the fatty acid oxidation cycle, oxidative phosphorylation, all the goodies, everything the mitochondrion did was systematically explored. [Grant, 6]

The feeling of heady excitement conveyed by this respondent was accompanied by confidence that any problems in laying bare the details of oxidative phosphorylation would soon be overcome.

> 2C
>
> It started when [Fennell] was a post-doctoral fellow and we ate lunch together every day. I had been working on an enzyme which was called glyceraldehyde phosphate dehydrogenase. That is, on substrate-level phosphorylation, it's a key enzyme in fermentation, which results in ATP, and I worked for a time on that enzyme. I showed that there was an intermediate, an oxidised intermediate, a thioester intermediate, which is formed during the action of the enzyme.
>
> Fennell, who was working on oxidative phosphorylation, said immediately, 'Well, why is that not also the mechanism of oxidative phosphorylation?', and so the model was built according to something which biochemists knew about and it was what we were used to. [Merrifield, 3]

Fennell formulated a theory to explain oxidative phosphorylation in terms of an unidentified 'high energy chemical intermediate' which, it was hypothesised, played an analogous role in oxidative phosphorylation to glyceraldehyde-3-phosphate dehydrogenase in substrate level phosphorylation. The latter enzyme was known to provide the chemical link between the oxidative and the phosphorylating reactions in substrate level phosphorylation, thus 'coupling' them together. Researchers turned to the search for the intermediate of oxidative phosphorylation.

> 2D
>
> They were quite convinced that electron transport was going to be just as easy a thing to sort out as glycolysis in solution. You had to have a series of carriers. Admittedly these things were not readily purified. But it became apparent that they reacted in an orderly sequence in the membrane. The game was – spot the missing factor. Everybody was convinced that you would just find the missing Fennell factor which would enable you to mimic all these effects in solution. [Whitehead, 5]

Unexpectedly, the search proved to be frustratingly difficult. The missing factor did not reveal itself.

> 2E
>
> I had the feeling that if you are a good boy, and do your homework, and study all the different parts of the mitochondrion, then in the end everything will fall into place, and all the mysteries will be resolved. It didn't turn out that way . . . We had some very exciting times, isolating these systems and proving the chemistry of it. But when it came to mechanism, a leap forward to the strategy of the structure, how was the mitochondrion designed – total failure, nobody succeeded. We were no better than anybody else. It was a great shock to me. [Pugh, 6]

2F

What I always say to undergraduates is: [Fennell's chemical intermediate hypothesis] looked very nice and you could forgive all of us working in this field for not finding [a chemical formula for the unknown intermediate], because that's rather like looking through a haystack for something without knowing what it is. A rather tall order. [Ashwood, 18]

Nevertheless, a succession of claims to have found the missing intermediate were made at intervals during the early 60s. In each case, the claims were eventually shown to be false.

One perspective on the state of research at that time is given in the following account by a scientist who was then a post-doctoral student at a prominent laboratory.

2G

I think at that time the general view in the field was that you could make ATP in a soluble system, so all you had to do was throw in . . . some cytochrome C [a component of the respiratory chain], take ADP and phosphate and then take [a proposed intermediate] enzyme, which made ATP. And at that time Perry had just isolated ATPase from beef heart mitochondria and Milner's lab had just isolated what they call an ADP/ATP exchange enzyme. And there were many papers, if you go back in that era, even showing that you could add this enzyme back to particles that were deficient in it and you could reconstitute ATP synthesis.

And so I worked on that system for the better part of three years . . . And the net result is that we could never really show any specificity of the enzyme in terms of its interaction with cytochrome C. Some other people had shown that it was specific for cytochrome C. And then Smith and Pugh and his collaborators had even indicated that they could take cytochrome C and basically the same enzyme and they would add ADP and phosphate and get reconstitution of oxidative phosphorylation. You had Milner's laboratory who wanted to do the same thing, and Perry's too as well . . .

It never occurred to me that people could say things like this, very important people, and turn out to be completely wrong, or perhaps had fabricated the whole thing. And I remember Perry went to Pugh's laboratory and tried to reproduce Smith's results, or even see the data, but apparently they couldn't come up with the data and it was learned that Smith had apparently fabricated the whole thing . . .

I thought well, the field is really screwed up . . . and I didn't believe half the stuff that was published, because I saw that we were dealing in this field with people who had very strong egos, who were trying to get an answer very rapidly, and who weren't cautious about what they were trying to do. [Carless, 4–6]

The incident with Smith was recalled by many of our respondents. The story goes that Smith had told his laboratory director that he had performed experiments which successfully identified an intermediate. The director announced this finding publicly at a conference, but later retracted the claim since Smith was unable to reproduce his results when the experiments were repeated under supervision.

It was during this period that Spencer formulated an alternative proposal which came to be known as the 'chemiosmotic hypothesis'. The first paper outlining his ideas in relation to oxidative phosphorylation was published in 1961, and this was followed by a series of publications developing the theory and providing some experimental support for it. The central ideas of the chemiosmotic hypothesis are that the creation of ATP takes place in the mitochondrial membrane; that the respiratory chain is located in the membrane and operates to divide hydrogen into electrons and protons; that specifiable numbers of protons are in effect transported across the membrane, thereby creating a gradient of protons and an electrical potential across the membrane; and that this proton gradient and difference in electrical potential, funnelled back across the membrane, provides the energy necessary to bind together ADP and phosphate.

Later, some researchers began to refer to parts of the respiratory chain as 'proton pumps' to emphasise their role in transporting protons across the membrane to create a proton gradient and electrical potential, although Spencer himself never seems to have used this term. Spencer's hypothesis avoided the need to postulate an unknown chemical intermediate, and explained why none had been discovered. But it was a somewhat unusual theory in the context of biochemistry at that time, because the proposed mechanism necessarily involved a membrane across which the proton and electron potential gradients were developed, and because Spencer's explanation drew on irreversible thermodynamics, a field in which most biochemists were not knowledgeable.

The initial reactions to Spencer's theory were not encouraging.

2H

When Spencer's paper first came out it was given very little attention really. There had been similar ideas in the literature before then – Robertson, and going back to Lundegardh, and so the field was familiar with the general idea that electron transport chains could act as proton pumps. Spencer did two new things: he suggested that . . . hydration and dehydration could lead to the pumping of protons . . . And his other new idea was to suggest that the sort of pump which Lundegardh had originally suggested . . . could be improved if you had an electron transport chain which is folded across the membrane with alternate hydrogen and electron carrier loops. That idea wasn't too clear in his first paper, but it certainly became the dominant idea in his later papers . . .

I think the major laboratories in the field were just not disposed to think of Spencer's hypothesis at all – I think that almost certainly someone in all those labs read the paper and put it to one side . . . I think that Spencer's paper was regarded as interesting but eccentric. [Richardson, 6–7]

Indeed, during the early 60s, his work was all but ignored, except by one British group of researchers who were interested in an issue which at that time seemed to be rather peripheral to the study of oxidative phosphorylation. These researchers were studying the action of 'uncouplers'. Oxidative phosphorylation involves two coupled reactions: respiration and the production of ATP. Normally these proceed hand in hand, with the energy produced by respiration consumed by ATP formation. However, certain reagents appear to uncouple them, so that, for instance, respiration continues rapidly, but without any accompanying formation of ATP. Many of the uncouplers were also known to be capable of transporting ions across the membrane. According to members of this group, researchers who were working from the basis of the chemical intermediate theory found it difficult to formulate convincing theories to explain why the uncouplers had an effect on oxidative phosphorylation.

The British researchers working on this topic picked up Spencer's work and used it to develop what they considered to be more adequate theories of uncoupler action. One of the members of this group related how he

21
first heard about the chemiosmotic theory as one always does these things, by word of mouth. A man with whom I was collaborating at the time . . . had met Spencer. Spencer had got a new theory, respiratory chain phosphorylation, and he's sitting there watching a pH meter needle move. He said, 'I can't see it move, but [Spencer] says it does', and he told me about him, and had a good laugh and decided Spencer was a bit mad; then the paper came out and I don't think at that stage it was taken very seriously. Now one of the problems is that we don't read enough outside our field and all Spencer's previous work had been in things like the *Journal of General Microbiology* and so on, and I don't think I'd ever read anything he'd written beforehand, and so I was not familiar in the least with his thinking. But we really only started to take things seriously when we started work on ion transport and then it became increasingly obvious that there was an economy in the chemiosmotic hypothesis describing what was going on which went right across the range of what we were doing . . . so that one became convinced that this really was more likely than the other thing [the chemical intermediate theory]. [Burridge, 9]

In the United States, however, the chemical intermediate theory still dominated the field:

> **2J**
> A lot of experimental results and valuable information was gained by the supposition that there were chemical entities in the respiratory chain [i.e. chemical intermediates] that were generated as electrons passed through – this led to Gowan's masterful analysis of the components of the respiratory chain, the cross-points and the change in the redox balance between carriers when you added ATP – all this was generated by the chemical theory, so they had no grounds for abandoning it. They regarded with some hostility any attempts to cut this from under their feet, they found it good and solid and their careers were based on it. [Roberts, 14]

Although this speaker suggests that the chemical intermediate theory was proving to have continuing value, and was generating 'masterful' analyses, other respondents painted a different picture of the situation in the United States at the time:

> **2K**
> The big Smith thing undermined a lot of confidence, not only in Smith and Pugh, but really in themselves. And I think there was a very big temptation at that time for people to get more slapdash rather than less because they were just desperate. They had to make the breakthrough and they were looking for shortcuts . . . You had people like me with no great ambitions to win Nobel prizes who were just prepared to work at it and just see what came out. You had the established people who in most cases were getting desperate, because they had committed, because they had made too big an investment. They were like in a game of poker where they had got all their money in the pot. They just had to keep upping the ante. And then there were a lot of very arrogant people who were coming in from other fields and saying, 'You chaps don't know a thing, I am a chemist, I don't think there is a problem at all' . . . And we had physicists coming in and saying the same thing and within about three years they were in just as bad a position as the old established people. So at that stage, well, it was exciting but also a bit of a mess. [Harding, 7–8]

An American scientist who was using the chemical intermediate framework at that time recalled his reactions to the chemiosmotic theory:

> **2L**
> You see the problem is that at first sight most people look at what he says and they just say, 'God, you can't understand it!', that's the first thing, especially in the late 60s, early 70s, you just didn't know what he was talking about. It was so much to buy, that here was this respiratory chain squirting out protons and making this magic electro-chemical proton. A biochemist is someone who is trained to purify, characterise and

crystallise. Nobody is going to purify, characterise and crystallise an electron gradient. [Cookson, 17]

Spencer travelled to the United States several times during the mid- to late 60s to speak at conferences. A senior US biochemist remarked that:

2M

American biochemists working on oxidative phosphorylation until late in the 1960s were very much committed to the chemical coupling hypothesis. Their whole outlook was that way . . . On three occasions [Spencer] came over. He didn't come over very well at first in this country at all . . . I still remember, and I had occasion not long ago to compare notes with other people who were present at that time, apart from a very few people he was greeted with complete disbelief and that even by some extremely intelligent people in this country. People just weren't thinking in those terms at all then among the American biochemical fraternity. And we'd had this period in which Pugh would every year or two make a pronouncement, 'Well, next year we'll have the problem all solved', and it never did get solved. [Milner, 9–10]

The British scientists working on uncouplers, who had already accepted chemiosmosis, also visited the States, and found the researchers there to be relatively ignorant about Spencer's ideas. A scientist who at the time was a student attached to that group said,

2N

I went over to work with Gowan . . . for a month. That was at the end of my first year as a research student. In [the laboratory he visited] there were quite a few people who were at the forefront of Oxidative Phosphorylation . . . I had to give a seminar there and that was really quite difficult because I had only done one year of my PhD, but actually I felt quite cocky about it, about half-way through the seminar, because I realised then that the people at this institute were lagging behind on these new ideas. And I was fairly confident saying, 'You are wrong here', and 'You are wrong here.' I was able to say that standing up in front of them because they hadn't given sufficient thought to Spencer's ideas. [Crosskey, 4–5]

In part as a response to these visits, and to the evidence produced both by Spencer himself and by his British supporters, some American scientists published work reporting results of experiments which, they argued, cast doubt on the chemiosmotic hypothesis. However, the supporters of chemiosmosis thought little of the quality of these experiments.

2O

There were a lot of experiments that were coming out at the time which were saying, 'This disproves the chemiosmotic hypothesis because . . .' and then giving some stupid reason. [Crosskey, 3]

2P

Until the 70s, at least in some of the major labs in the field, nobody had made the effort to understand the chemiosmotic hypothesis. It's pretty obvious from the kind of things that they were saying at conferences or in papers, and the number of people who 'disproved' the chemiosmotic hypothesis that they were just making trivial physical-chemical mistakes, this sort of thing, or were interpreting experiments in a way which was plausible, but not probable, in the sense that they would fail to take account of other possible effects. [Richardson, II.1]

During this same period, two other theories to explain oxidative phosphorylation were proposed by American biochemists. Watson's 'conformational coupling' hypothesis, formulated in 1964, suggested that the energy necessary for the formation of ATP was not stored as a chemical intermediate or as an electro-chemical gradient, but as changes in the conformation or shape of the molecules in the mitochondrial membrane. Pugh proposed a theory in which the coupling between the respiratory chain and ATP synthesis was conceived in terms of mechanical movements of the membrane. Neither of these theories became particularly popular, and comments made by our respondents suggested that, in retrospect, they were either dismissed entirely, or were regarded as variations on the chemiosmotic and chemical intermediate theories. The following passage from a scientist who has always thought the chemiosmotic theory to be correct, is typical of the reactions that these two theories received from many of our respondents:

2Q

[Pugh] had a chemist drafted in to prove the conformational hypothesis, which Pugh had jumped upon and used all his electron microscopes and things to 'prove'... He was an ace chemist. Very, very good. He went to work there for about a year. At the time they published this chemico-mechano-chemiosmotic six-stage theory or whatever. In which all you had to do was take out four stages and you could throw away the rest. This guy went and listened to Pugh and after he'd been there about six months, he came over to see me. And he said, 'Now look, I hear you know all about the chemiosmotic theory.' So we had a long chat and we talked about it. He went away again and came back three months later and said, 'I'm convinced the chemiosmotic theory is probably right', he said. 'But Pugh doesn't believe me.' So a month after that he was gone. [Barton, 15]

Meanwhile, there were continued attempts by some to find experimental proof that the chemiosmotic theory was incorrect. The following passage describes one such experiment. The chemiosmotic theory included the postulate that oxidative phosphorylation would only take place in closed membranes. If the membrane were ruptured, for example by exposing it to intense sound waves, the theory predicted that phosphorylation would not be observed. In the experiment described, as in many others of this kind, the membranes were in the form of artificially produced 'vesicles' rather than intact mitochondria.

> **2R**
>
> There was a paper ... in a meeting about 1970 ... They'd sonicated these things for about thirty minutes – a desperate attempt. They implied, maybe they were more explicit and said that they'd got oxidative phosphorylation in open vesicles. That was the point of the experiment, anyway. I remember in [another] meeting in 1972, somebody had written an abstract saying that they had demonstrated oxidative phosphoryla-tion in a membrane-free system derived from a bacterium ... Normally these ten-minute papers, not many people attend. But I noticed that the room was filled, and the usual anti-chemiosmotic gang were all there like vultures. But the evidence that there were no membranes there wasn't very satisfactory. You could see them going away a little disappointed. Once again, this horse hadn't run very far. [Aldridge, 8–9]

On the other hand, two experiments, one carried out in 1966 and the other over a period of years in the late 60s, had some influence in Spencer's favour. These were Miller's 'acid-bath' experiment, and Perry's recon-stitution experiment. The former produced results which were hard to explain in terms other than those of the chemiosmotic theory, but was an experiment on photo-phosphorylation in plants, rather than on oxidative phosphorylation. The second experiment involved artificially constructing a working phosphorylating system from separate components taken from a number of different sources. It was considered that if the chemical intermediate theory were true, the resulting reconstituted system would be most unlikely to form ATP, whilst the chemiosmotic theory predicted that such a system could be made to work. The experiment showed that small amounts of ATP were formed.

> **2S**
>
> The reconstitution experiments were just beginning. I think Perry and Czigler, that is perhaps the touchstone, that is the one single experiment that people cite – you take the ATPase away from the respiratory chain and it still behaves like a proton pump and you cannot postulate some hidden lost proton pump stuck in between the respiratory chain and the ATPase which is what the chemical people were driven to. [Roberts, 17]

Gradually, however, the tide seemed to turn in favour of Spencer. A number of respondents put this movement down to the influence of younger scientists entering the field without the 'prejudices' and preconceptions of the older ones.

2 T

When I came here I knew nothing. All the seminars they had here were just incomprehensible, but it was clear that everybody was against Spencer heavily at that time and he wasn't doing too well. There was a lot of 'evidence' against his ideas. The movement towards accepting it . . . was due to a lot of young people coming in. People who were very strongly involved with Spencer. The thing about [the director of the laboratory] was that in spite of his anti-dogma thoughts, he would always invite people here who opposed him and that was one of his terrific characteristics . . . and so a lot of people [came], particularly young English people, the national support club of Spencer. [Jeffery, 20]

Thus, by the early to mid-70s, the big American laboratories had begun to become more friendly to the chemiosmotic hypothesis, although there is some disagreement about the extent and timing of this shift. For instance, Spencer wrote in 1976 that, in his opinion,

2 U

Several of the more eminent authorities in the field of oxidative phosphorylation are reluctant to agree that coupling between the proton-translocating respiratory chain system and the proton-translocating ATPase system, plugged through the coupling membrane, is due to the proton current circulating between and vectorially through them . . . they have preferred to believe, in keeping with the traditionally scalar origins of their conception of metabolism, that coupling is achieved by some unidentified energy-rich chemical intermediates or by some direct interactions between components of the respiratory chain and reversible ATPase systems.

In contrast, Norton, speaking of the same period, noted that,

2 V

There was a substantial shift about the time I went to the States. [Earlier], it was unthinkable that Fennell should talk in chemiosmotic terms. And just as I was leaving [for the States], he started to. I think that made a big impresssion on a lot of people. Their hero, the chemical theory's hero, starting to switch positions, or at least to admit that it was very *reasonable* to talk in chemiosmotic terms. And Gowan as well, about that time, finally got worn down by the stream of British post-docs coming into his lab and telling him chemiosmotic things· and thinking chemiosmotically. But the respectable way to talk about things has become chemiosmotic and the odd way non-chemiosmotic, rather than the other way round. That switch was occurring while I was there. It was surprisingly rapid, unexpectedly rapid. [Norton 34–5]

Whilst according to most respondents, the tide was clearly flowing towards the adoption of chemiosmosis, there was continuing controversy about some aspects of the theory. One issue that remained unresolved was that of 'stoichiometry'. Respondents used this term to refer to the number of protons that are transported across the membrane as one electron passes down the respiratory chain:

> 2W
> There's an interesting thing that has happened in the last five years and that is that five or ten years ago people were trying to show that either the energy balanced, or it didn't balance. These two numbers, delta P is the size of the proton gradient and delta G_p is the energy needed to maintain ATP. People wanted to show that delta P and delta G_p balanced and they used Spencer's numbers. Spencer predicted certain stoichiometries between the number of electrons, the number of protons and the number of ATP molecules. So, in those days we used his numbers and applied those numbers to the data and the thing roughly balanced, but there's definitely been a swing away from that in the last few years. The assumption has been that the chemiosmotic hypothesis is right and the energies will balance, but what numbers do we have to use? What stoichiometries do we have to use to get them to balance? . . . So there's a lot of argument as to whether 'n' is 2 or 3 or 4, whereas ten years ago we were assuming Spencer says 'n' is 2, we will take 2 and plug it in and roughly it balances. And any imbalance we put down to, well, a data problem, error. [Crosskey, 9]

As this speaker notes, Spencer's chemiosmotic hypothesis assumed a stoichiometry of 2. Moreover, Spencer subsequently formulated a fairly elaborate biochemical mechanism, involving proton-transporting pathways that formed so-called 'loops', based on this stoichiometry. However, in the late 70s, doubts were cast on this figure; experimental evidence was interpreted to support stoichiometries of 3 or, in some cases, 4. But to most respondents, the stoichiometry question was a 'side issue', a detail which had only become important because the overall chemiosmotic framework was no longer controversial, and because Spencer seemed to some participants to be unreasonably inflexible in refusing to accept that his numbers were not the correct ones.

At the time we interviewed our respondents, in 1978–79, the controversy over stoichiometry had not been settled. Nevertheless, in the autumn of 1978, Spencer was awarded the Nobel Prize for his contribution to the understanding of biological energy transfer. Some respondents seemed a little surprised that Spencer had been given the Nobel Prize, since they regarded his theory as still subject to doubt. Although as one speaker said,

2X

The chemical [intermediate] theory is as relevant as phlogiston. Only in textbooks is there any discussion about conflicting theories. [The chemiosmotic theory] has taken over. [Aldridge, 29]

others were less certain that chemiosmosis, as formulated by Spencer, had 'taken over'. Most respondents accepted that the field, in general, now concurred about the existence of a pH gradient and a membrane potential that could be used to drive ATP synthesis, but few would accept that there was such general agreement about more specific details. The following is characteristic of respondents' present-day views about chemiosmosis, although different speakers emphasised different points of contention, doubt or agreement.

2Y

Spencer's proposals can be broken down into three parts. The main hypothesis, which is that all the energy which is conserved during electron transport passes through an electro-chemical gradient on its way to different functions to drive ATP synthesis and transport. That's the idea he was awarded the Nobel Prize for, I believe. Then, how is that electro-chemical gradient, the proton gradient, formed and used. How the gradient is formed, the idea of loops, some people have subscribed to them, but more and more people seem to be thinking that there's a proton pump, rather than the actual oxidation-reduction reaction involving a hydrogen ion donor and an electron acceptor . . . that's Spencer's loop idea. More and more people seem to be coming round to the pump idea . . . The mechanism of ATP synthesis, when he takes this electro-chemical gradient and says the protons on the outside can rip an oxygen molecule off phosphates so that it spontaneously combines with ATP, that is I suppose an attempt to use the gradient directly without any intermediate step or conformation step or chemical step. The idea is to try to make ATP directly from a proton gradient. It just doesn't agree with a tremendous amount of data that's in the literature, it's contradictory. [Hargreaves, 21–2]

Thus, this speaker, as soon as he begins to talk about specifics, and in particular about ATP synthesis, the area he is working on, starts to identify what he sees as serious shortcomings in Spencer's ideas. Others suggested that Spencer's theories, although previously regarded as too complicated to be worth pursuing, were in fact too simple to represent the facts correctly, and that combinations of Spencer's ideas with elements of, for instance, Watson's conformational coupling theory were likely to be needed. These scientists seemed to think that much further research to develop more detailed and refined theories was required. In contrast, a few of those we spoke to suggested that knowledge about oxidative phosphorylation was now so advanced that there was no longer anything interesting left for them to work on, and that they were therefore moving out of bioenergetics into other areas of biochemistry.

Any attempt to draw conclusions about the present level of agreement about the chemiosmotic hypothesis within bioenergetics is further complicated by the fact that different respondents describe the theory in quite different ways. It is far from clear, therefore, whether all those who say that they accept the hypothesis are subscribing to the same theory. For instance, compare the following rendering of the chemiosmotic hypothesis with those cited above:

> 2Z
> The field changed enormously because of Spencer. It was during this period of time when I came in 1964 that everybody was thinking of chemical mechanisms for how ATP would be made in a soluble system. And it was only then, at that stage of the game, that Spencer came out with his early papers in *Nature*, and then finally those two little books that he sent me, in which he emphasised that all biological systems that carry out the synthesis of ATP are membrane systems and they are all closed systems and that nobody had ever isolated a chemical intermediate before and so maybe they didn't exist. Which was a little far-fetched, because you've got to have chemical intermediates to make ATP. But in a way he had chemical intermediates, he was simply stating that there was a different way of making ATP than having a high-energy chemical compound of the electron transport chain. I think that was his main emphasis, to do away with any high-energy intermediate which was associated with electron transport. [Carless, 23]

The speaker describes the chemiosmotic theory here in a way which emphasises the close connection between his version of it and the fundamental ideas of the chemical intermediate theory, the theory that he had earlier espoused. He states, for example, that even Spencer's theory had chemical intermediates 'in a way'. His implication that the chemical and chemiosmotic theories are not really so very different contrasts markedly with the opinions of other respondents (such as those quoted in 2H, 2L, and 2R). We shall be returning to consider in more detail the topic of agreement and consensus in chapter six.

Alternative versions of the history

In order to give a view of the development of the field as a whole, we constructed the history of oxidative phosphorylation presented in the previous section from a patchwork of quotations from many different respondents. Overall, it is an account which we believe many participants would recognise and accept as a reasonable approximation to the way in which the field developed. But it would be quite misleading to suggest that this history describes 'the way things really happened'. There are several reasons for this. First, the history was created from portions extracted from interviews, in which speakers were asked to give an account of their scientific biographies, and to comment on the history of the field from their own knowledge. This means that many respondents gave us an account in which their own activities were the major focus, and which was therefore inevitably a very partial account of the development of the field as a whole. It cannot be guaranteed that even when all the speakers' accounts have been put together, every significant event will have been included. Secondly, the speakers gave us retrospective accounts, in which, to an unknown degree, the events they reported may have been 'reconstructed' in order, for instance, to fill gaps in the speaker's recollection. Thirdly, the historical events that were mentioned in each speaker's account were given meaning in relation to particular 'dramatic episodes' that varied between speakers and during the course of each interview. Thus some respondents mentioned particular events and others did not; different speakers emphasised the importance of specific events quite differently; and whilst some speakers frequently talked in general terms about the development of the area, others focused more narrowly on their own involvement in it. In short, the quotations have been taken out of their particular contexts.

These 'problems' with the data are not, however, confined to extracts from interviews. They would not be avoided by recourse to 'harder' data, such as, for example, laboratory notebooks, contemporary publications, diaries and so on. Descriptions of events from these sources would also be subject to precisely the same 'problems'. Particular examples of such data could not be expected to provide more than partial descriptions of events, thus still requiring the historian to construct a coherent, overall view from a patchwork of items of data. Furthermore, even items of written data are reconstructions of past events, produced by the author for some particular context, and oriented to that context. For example, as we shall show in the next chapter, the introductory sections of research papers, which often present reviews of prior work and which to that extent are sources of historical data, can be seen to be rather finely crafted reconstructions in which certain kinds of events and actions are systematically excluded.

Thus, although we recognise that the history we constructed is but one possible version of the history of the field, this 'weakness' cannot be remedied by relying instead on other kinds of data. Other versions of the history will remain both possible and plausible. Examples of such other versions may be found in our interview transcripts and it is to these that we shall now turn.

From the very large variety of 'events' which *could* have been mentioned in outlining an historical sequence, our respondents inevitably chose only a small number to talk about. One of the ways in which such selections appear to be made in the course of conversation is to choose those events which, when put together, compose a dramatic story. Thus speakers, asked to describe their scientific biographies, did not merely list the papers they had published, the research problems they had worked on, or the laboratories they had worked in. Instead, they 'told the story of their lives', connecting the events into a dramatic narrative.

Examining the narratives that were offered to us, many, but not all, the respondents used a story line which could be called the 'David and Goliath' theme. This theme also enters the version of the history we have presented in the previous section. In brief, the story line runs like this: The field was dominated by three very large, very well funded US laboratories. The directors of these laboratories were arrogant, over-confident, and determined to be the first to crack the problem of oxidative phosphorylation. The competition between them was so intense that their assistants were pressurised into making fraudulent claims, bitter disputes emerged at scientific meetings, and the field as a whole began to have a bad reputation. But 'David', the scientist from Britain, launched an attack on these 'Goliaths' single-handed, and after battling for nearly twenty years against them, and resisting all their attempts to defeat him, cracked the problem and won the Nobel Prize, to the Goliaths' fury.

The following passage, rather longer than those we have previously quoted, taken from one respondent's interview transcript, will serve to illustrate how the historical accounts which we have used above are often told as a moral tale, in this instance, of David versus Goliath.

> **2AA**
> The personalities in the field were very dominant. The big groups were dominated by the leaders and there you had Perry, Pugh, Gowan, and then to a lesser extent . . . Fennell, who happens to have a reasonable group, but regarded himself as a sort of grand referee in the field. So every meeting you went to, you had these personalities always pronouncing the gospel. In the US all of this was dominated by those particular groups . . . I don't know when Spencer first went and gave those Federation meeting talks, '66 or '67. I think people at the time didn't take him very seriously, but I think he attracted a good deal of attention among, first of all, the chloroplast people because they were moving that way already. But he also attracted in the US, more particularly, support among the bacterial people . . . But still a lot of dubiousness among the mitochondrial people. Whereas over in Britain, the reverse was true. I think he was beginning to attract a strong following in Britain particularly, which overflowed to the continent, but it took time . . . I don't think *he* helped . . . People didn't

really believe it and he was advocating what *he* thought was a reasonable hypothesis, which wasn't going over. But he wasn't dealing with reasonable people. The opposition was in some cases quite vitriolic . . . It was the norm in the field at the time. Particularly in the 60s, the oxidative phosphorylation field had the reputation that if you went to a Federation meeting, all the meetings were crowded because everybody went along because they knew there would be a damned good fight there . . . I think it basically relates to the fact that progress was so slow that the points that were being discussed were so nit-picking and the field wasn't moving at all. Large, huge, effort went into it in many laboratories. People have put their lives into this field and with huge groups associated with them, it's been very disappointing in this respect. [Peck, 21–3]

In this tale, the leaders of the major US laboratories are described in critical terms as 'dominant', 'unreasonable' and 'pronouncing the gospel'. In contrast, Spencer is simply engaged in advocating what he thought was a 'reasonable hypothesis'. The opposition to Spencer is indicated to consist of 'groups', 'laboratories', or even 'huge groups' that are dominated by their leaders. In contrast, Spencer attracts a strong following among presumably independent researchers. Spencer himself is referred to throughout the passage as a single actor, a lone person without the support of great resources. He had to contend with people whose unreasoning opposition was 'vitriolic'. Yet these people were making no progress; the discussion in meetings was 'nit-picking'. Furthermore, their failure to accept Spencer's theory is contrasted with the much more ready acceptance of those working with chloroplasts and bacteria, areas which were presumably not dominated by leaders having the opprobrious attributes characteristic of US research on mitochondria.

This kind of moral tale features in a number of accounts, although not all used the same David and Goliath theme. A common, but less popular, theme was the one in which a researcher, often the speaker himself or a colleague, laboratory director or other mentor, produced a masterful theoretical or experimental advance which failed to get proper credit from other scientists. When respondents tell their story in these terms, the history of the field appears rather differently. The following passage illustrates this second theme.

2 A B
I happened to run into one of Pugh's papers . . . this was in the late 60s, I think. He was developing some ideas about what he called an electro-mechanical model. And it seemed that somebody with a training in solid state theory could *do* something there. And he was rather unique in the field, in the sense that he was trying to construct physical models where other people were not trying to do that. I read more about the field;

in fact I don't think I could have penetrated the field without his papers, because everything was just a giant mess and he was the one who was trying to create some order in it . . . I started reading his papers because I couldn't make head nor tail of the chaos of the mostly bio-chemical type of papers and it was his organisation of the material which allowed me to penetrate into this field. We started collaborating and evolved this model . . . The way I perceived it at the beginning was that there were several experiments for which the details were not clear, but some kind of picture *had* to emerge and as far as I could see he was the only one who tried . . . I felt at the time that he developed a model which was very sound. There was no proof for it; there was no disproof for it either . . .

The chemical theory didn't seem to *say* very much. The chemiosmotic theory I always felt was weak in one respect, which was that the proton gradient develops across the membrane and there is an equilibration with the two liquid phases before the protons would transmit the energy and make ATP. Which was something that Jennings had already emphasised: he came out with the proton-in-the-membrane hypothesis about the same time as Spencer and I thought that that was a much more *reasonable* hypothesis. However, what I felt was lacking was a physical mechanism by which the energy is transmitted, and [the speaker's] theory was supposed to supply a possible mechanism for such energy transfer.

But looking back on it now, I think it would have been better never to have got into [the area]. Because the time when we got into trying to develop such a model was the time when what I would call the first generation's theories came to an end . . . It is models like the chemiosmotic model, the chemical hypothesis, etc., which try to make some simple pictures, they just about ran out of usefulness at that time. We picked up on the last edition to this first generation's theories. My feeling is that just about that time, much more detailed experiments were being produced, which as far as I know have led to *no* theoretical understanding at all. But what this means to me is that it is time to keep quiet and wait for the second-generation experiments which are of greater detail and precision. I think most of these experiments are more confusing than helpful at the moment, but it also means that the precision of the theory is way behind the precision of the experiments at the moment. [Hinton, 2–3, 6]

In quotation 2Q we saw a speaker dismissing Pugh's model, formulated in the 60s, as merely a version of chemiosmosis. The present speaker, however, reports that at that time he saw this model as 'trying to create some order' from 'a giant mess'. Pugh is cited as trying, as nobody else, to construct physical models to reduce the 'chaos'. Ultimately, although the model he developed was 'very sound', he was defeated by the fact that he was at the tail end of the 'first generation' of theories. It is clear that the overall characterisation of the research field is treated quite differently by this speaker as compared, for instance, with the respondent quoted in the previous passage, 2AA. The chemiosmotic theory is here considered to be merely one from a number that 'try to make simple pictures', and which together constitute the 'first generation' of inevitably inadequate hypotheses. Real progress will have to await 'second-generation' experiments. There is no mention of Spencer having a heroic role in the development of the area. Moreover, Spencer's theory is not seen as having 'solved' the problems of the field; indeed, it seems to be suggested that the problems themselves are not yet clearly defined, and await clarification by 'second-generation experiments'.

The two researchers whom we have quoted in this section had both contributed to oxidative phosphorylation research. Both were describing events which occurred during the mid-60s. Both were fairly young, but were established scientists by the late 60s. Furthermore, both scientists had had close contact with the same laboratory director, whose interests and ideas had greatly influenced their subsequent research. Despite these common features in their biographies, the historical details they offered to us and the perspectives they took were so different that it is quite easy to believe at first sight that they were not describing the same research area. Many further examples of variability can be found by comparing other accounts obtained from our respondents, and as we argued in the first chapter, this variability casts considerable doubt on the worth of traditional forms of analysis that attempt to reconcile these variations.

As we showed in the previous section, however, the existence of this variability is not, in itself, a bar to the construction of a version of history by the analyst. In the main body of this chapter we have compiled our own 'folk history' of 'ox phos' in order to introduce some of the terms and issues to which participants will constantly refer in subsequent chapters. We have also illustrated the kind of recurrent variation in accounts which is a feature of our data. However, in the interests of making the scientific issues clear, we have merely pointed to, but not systematically described or analysed, such variation. In the next chapter, we start to show that a detailed examination of variations in participants' discourse reveals regularities of some sociological significance.

3

‥‥‥

Contexts of scientific discourse

In chapter one, we argued that interpretative variability within participants' discourse is sociologically important. In the second chapter, we illustrated this variability in relation to a couple of accounts of the development of research into oxidative phosphorylation. In this chapter, we will begin to show that participants' discourse, although varied, displays certain observable patterns. We will examine two contexts of linguistic production, namely, the experimental research paper and the semi-structured interview involving biochemists and non-biochemists. We will show that participants' accounts of action and belief are systematically different in these two settings. This will enable us to identify two major interpretative repertoires, or linguistic registers, which occur repeatedly in scientific discourse.

It is important to emphasise that when we use the phrase 'social context', we are not referring to phenomena which exist independently of participants' discourse. For social contexts are themselves *products* of discourse. It is through their recurrent patterns of language-use that participants construct such phenomena as the 'formal research literature' and 'informal' interaction. To describe the social context of the formal research literature as being distinct from the context of informal interaction is essentially to refer to systematic differences in the ways in which scientists construe their actions and beliefs. Phrases like 'the formal research literature' are labels employed by participants as well as analysts to distinguish variations in the forms of discourse by means of which members construct the meaning of action and belief.

Halliday makes this point more precisely as follows.

> Let us assume that the social system (or the 'culture') can be represented as a construction of meanings – as a semiotic system. The meanings that constitute the social system are exchanged through a variety of modes or channels, of which language is one ... Given this social-semiotic perspective, a *social context* ... is a temporary construct or instantiation of meanings from the social system ... [the] components of the context are systematically related to the components of the semantic system; and ... given that the context is a semiotic construct, this relation can be seen

as one of realisation. *The meanings that constitute the social context are realised through selections in the meaning potential of language.*[1]

For the purposes of our analysis, reference to the existence of different social contexts in science is simply another way of drawing attention to patterned variations in the discourse through which scientists construct their social world.

In this chapter, then, we will begin to describe some of the systematic ways in which scientists select from the full meaning potential of their language as they construct formal and informal contexts within the scientific domain. We will illustrate how they draw selectively on two interpretative repertoires as they depict action and belief in ways which are appropriate to the different interpretative contexts they are involved in reproducing. More specifically, we shall show that when scientists write experimental research papers, they make their results meaningful by linking them to explicit accounts of social action and belief; that the accounts of action and belief presented in the formal research literature employ only one of the repertoires of social accounting used by scientists informally; that these formal accounts are couched in terms of an empiricist representation of scientific action; that this empiricist repertoire exists alongside an alternative interpretative resource, which we have called the contingent repertoire; that this latter repertoire tends to be excluded from the realm of formal discourse; and that the existence of these formally incompatible repertoires helps us to begin to understand the recurrent appearance of interpretative inconsistency in scientists' discourse.

Full-length experimental papers are usually divided into separate sections: the Abstract, Introduction, Methods and Materials, Results, and Discussion. Although we believe that all these sections involve some kind of social accounting, we will concentrate upon Introductions and Methods and Materials sections. We will first of all present some passages from research papers to show that they do contain accounts of their authors' actions and to draw attention to the way in which these actions are presented. We will then compare these formal accounts with those given informally in interviews by the same scientists. This comparison enables us to demonstrate that the two kinds of account differ dramatically in several respects.

We will concentrate on material from just two interviews and on one published paper by each interviewee. We have chosen to proceed by means of a comparatively detailed examination of two papers, rather than by means of quantitative analysis of large numbers of texts, because it seems to us that in this way we can demonstrate most forcibly the capacity of particular scientists to produce radically different versions of given actions. Nevertheless, evidence will be required from many more research papers and from other research areas in order to establish any degree of generality for our conclusions. Fortunately, the relatively few previous studies of formal scientific texts in other areas have all furnished observation very similar to our own.

The points we intend to make below do not involve any appreciation of the technicalities of biochemistry beyond those introduced in the previous chapter. Non-biochemists should not be deterred, therefore, if they fail to understand completely the quotation from an introduction to a research paper which begins the next section. We use only four quotations with such a high level of technical content in this chapter, and only two of them need to be understood by the reader in any detail. Both these quotations are followed immediately by a 'layman's gloss'.

Social accounting in introductions

INTRODUCTION I (main author Leman)
A long held assumption concerning oxidative phosphorylation has been that the energy available from oxidation-reduction reactions is used to drive the formation of the terminal covalent anhydride bond in ATP. Contrary to this view, recent results from several laboratories suggest that energy is used primarily to promote the binding of ADP and phosphate in a catalytically competent mode (1) and to facilitate the release of bound ATP (2,3). In this model, bound ATP forms at the catalytic site from bound ADP and phosphate with little change in free energy.

A critical test of this proposal would be to measure energy-dependent changes in binding affinities at the catalytic site for adenine nucleotides. However, such measurements are complicated by the fact that mitochondrial membranes have numerous binding sites . . . An inhibitor that specifically prevents substrate binding at the catalytic site would prove very useful since it would allow binding events directly involved in catalysis to be distinguished from other processes that require bound adenine nucleotide . . .

An indication that the new phosphorylation inhibitor, efrapeptin, might bind at the catalytic site comes from studies with aurovertin . . .

In this paper, we report the results of studies on the mode of inhibition of oxidative phosphorylation by efrapeptin . . . It is difficult to accommodate these results in a single mechanistic scheme involving a single independent catalytic site for ATP synthesis and hydrolysis. As will be discussed, the data are more easily interpreted in terms of a multiple interacting site model, such as the one recently proposed by Bradshaw, Willow and Stein.

LAYMAN'S GLOSS I

ATP is one of a class of complex molecules called nucleotides. It is biologically important because it is a major source of energy in living organisms. ATP is formed by the combination of ADP and inorganic phosphate. The overall process whereby ATP is formed is called oxidative phosphorylation. The energy required for its formation (or catalysis or synthesis) is thought to be generated through a series of linked oxidation-reduction reactions. The point made by this author in the first paragraph is that, whereas many biochemists have believed that the energy produced by these oxidation-reduction reactions is used to bind together ADP and phosphate, there is now good evidence showing that this binding occurs at specific sites with little expenditure of energy. He refers to a new model of oxidative phosphorylation in which the free energy is used, not to *make* ATP, but to *release* it for physiological purposes.

He then goes on to state that this model could be tested by measuring the relevant binding affinities. However, this task is complicated by the fact that the membranes of mitochondria, which are complex intra-cellular particles within which these processes occur, have numerous binding sites in addition to that where ATP is formed (the catalytic site). Efrapeptin is then identified as a substance which appears to act only on the catalytic binding site and which should help the experimenter to make observation of that site alone. Finally, the author claims that the results obtained with efrapeptin are inconsistent with older views of ATP formation and are best interpreted in terms of an extended version of the new model mentioned in the first paragraph.

Many commentators have drawn attention to the way in which scientific papers are written in an impersonal style, with overt references to the actions, choices and judgments of their authors being kept to a minimum. In this respect introduction I, about half of which is reproduced above, is typical of scientific writing. Although three other scientists are referred to by name in the full text, there is in every instance a rapid return to less personal formulations and the authors themselves only appear once through their use of the pronoun 'we'. At various points in the exposition verbs usually associated with human agency are employed, but are often combined with some non-human 'agent'. Thus 'recent results' are said to 'suggest' certain possibilities, and 'studies with aurovertin' are said to 'indicate' others. Despite this impersonal style, which minimises explicit mention of social actors and their beliefs, it is clear that parts of the text implicitly offer accounts of the actions and beliefs of the authors and of their specialised research community. To this extent the introduction has a definite, albeit partly obscured, social component. The significance of the authors' findings is established at least partly by the way in which the social element in the text is presented.

The opening sentence, for example, is not a statement about the physical world, but about the customary nature of certain beliefs among a number of biochemists. This sentence could equally well have been written by a sociologist, trying to construct an interpretative analysis of social action in a research network. This similarity exists because the sentence *is* part of a subtle and organised social analysis. The beliefs in question are presented in a way which enables the authors to contrast them unfavourably with those of another group of scientists, to which the authors themselves belong. What is particularly noticeable about the first sentence is how the beliefs which it summarises are prepared for immediate rejection. Thus, instead of presenting the central idea as a reasonable, though inconclusive, interpretation associated with at least some experimental evidence, it is depicted in the text as mere assumption. Furthermore, no supporting literature is cited. The impression is conveyed that, although this idea may have been around for a long time, it has no firm scientific foundation and is not to be taken seriously.

The nature of the opening sentence prepares us to expect and to welcome the contrasting view which the second sentence reveals. Clearly the reader, as a scientist, is expected not to favour unsupported assumptions, but only views based on hard data. Consequently, the second sentence informs us that it is *experimental results* which suggest a significantly different state of affairs from that previously assumed. Reference is made implicitly to particular actors, when the phrase 'several laboratories' is used. But the authors do not formulate their argument, as they could have done with at least equal accuracy, in terms of two or more groups of scientists producing different experiments along with plausible yet differing interpretations of those experiments, but in terms of one group's *results* undermining the other group's *assumptions*. Although in sentence 2 the conclusions deriving from these results are presented simply as suggestions, they are described more strongly in the third sentence as constituting a model; that is, a systematic explanatory scheme with, as the next paragraph makes clear, a central proposition which can be put to the test. References are given for this model, in case the reader wishes to check its content or its empirical support. Thus, in the course of three or four sentences, the text has conveyed a strong impression, at least for readers unfamiliar with the topic, that the rest of the paper is based upon a well established analytical position which constitutes a major advance on prior work. This has been achieved, not by the presentation of biochemical findings, but by the characterisation of scientific action and belief within the authors' social network.

The formal account of collective belief offered by these authors would have been treated as misleading by many of the scientists we interviewed.

For many of them, at least informally, were highly critical of the model advanced in this paper and in particular expressed dissatisfaction with its supporters' failure to provide empirical clarification of the physical mechanisms involved. We put this point to the senior author during the interview.

> 3 A
>
> *Interviewer:* The most frequent criticism of the idea of conformational coupling that people have talked about is that it doesn't tell you anything about mechanism. How would you respond to that comment?
> *Leman:* I'd say they were right. We always feel a little embarrassed when we talk about conformational change. Because it's a vague sort of thing. But I think it is an important idea. The data seems to indicate that you need energy to release ATP from the enzyme ... I agree that it is aesthetically unpleasing not to have a very detailed account of what's happening. If it *is* an energy-driven conformational change, that change will never, not in our lifetime anyway, be described in very discrete steps. [29–30]

In this passage Leman qualifies his formal description of the merits of the model he is advocating, in response to our version of the informal comments of other researchers. This shows clearly, not only that other scientists would probably have introduced these results quite differently, but also that a quite different account could have been given by the authors themselves of the state of scientific belief within their social network.

In order to show that the research paper considered so far is not unique, let us look at another introduction.

> **INTRODUCTION II (main author Spender)**
> The chemiosmotic hypothesis (1) proposed, *inter alia*, that each span of mitochondrial respiratory carriers and enzymes covering a so-called energy-conservation site (2) is so arranged that $2H^+$ are translocated across the mitochondrial inner membrane for each pair of reducing equivalents transferred across that span. Evidence in favour of this value of 2.0 for the ratio of protons translocated to reducing-equivalent pairs transferred (i.e $\rightarrow H^+/2e^-$ ratio) has come mainly from one type of experiment. In this, the length of the respiratory chain under study has been altered by changing either the oxidant or the substrate (3,4).
>
> In the present paper we describe an independent method for measurement of the $\rightarrow H^+/2e^-$ ratio per energy-conservation site. The same substrate (intramitochondrial NADH) and oxidant (oxygen) is used throughout, but the number of energy-conservation sites is varied from one to three by using mitochondria from variants of [a particular yeast] with modified respiratory chains. We conclude that the $\rightarrow H^+/2e^-$ ratio is 2.0 per energy-conservation site.

LAYMAN'S GLOSS II

This paper is concerned with the series of oxidation-reduction reactions which are believed to occur in the membranes of mitochondria. This series of reactions is referred to as the respiratory chain. It is the respiratory chain which is taken to generate the free energy required for the formation and/or release of ATP. The author takes it for granted that this energy is furnished by a gradient of protons (H^+) which is created across the mitochondrial membrane by the action of the respiratory chain. He also takes it for granted that protons are carried across the membrane by pairs of electrons ($2e^-$) at three sites. The issue which he addresses is, 'How many protons are carried across at each site by each pair of electrons, i.e. what is the $\rightarrow H^+/2e^-$ ratio?'

In the first paragraph, he states that a figure of 2.0 has been obtained previously by means of experiments in which the number of sites in the chain has been varied by changing the substrate (the reagent which donates the protons and electrons to the chain) or the oxidant (the reagent which receives the electrons after the protons have been transported). This is possible because some substrates and oxidants operate at different points in the chain.

In the second paragraph he states that he has also obtained a ratio of 2.0 per site, using the same substrate and oxidant, but employing mitochondria from three different strains of yeast with respiratory chains of varying length. Later in the paper he states that these mitochondria have chains with either one site, two sites or the full three sites.

This introduction seems straightforward and unproblematic. A quantitative aspect of a major hypothesis is first identified. The authors point out that, although this part of the hypothesis has been experimentally confirmed, only one kind of experimental design has been employed. An alternative technique is then briefly described. And the introduction ends with a statement that this new technique produces the same results as previous experiments and therefore provides further support for the hypothesis. In what sense does this passage involve social accounting? In the first place, an account is being offered of the state of belief among those scientists concerned with the proton/electron ratio. Although it is not stated explicitly, it seems to be implied that there are no negative experimental findings which need to be considered and little, if any, disagreement about the scientific meaning of previous findings. Indeed, it is this form of presentation which enables the authors to depict their own results as primarily a contribution to experimental technique: 'In the present paper we describe an independent method for measurement of' the relevant ratios. Their actual observations can be treated as unproblematic, because they are portrayed as merely confirming what competent researchers already know. As a result, attention is directed to the novel techniques used to obtain these expected observations. In this way the authors' contribution to knowledge, and thereby the meaning of their work in the laboratory, is established by the manner in which the existing state of belief about this ratio is construed.

The content of this introduction differs considerably from the discussion of the paper in the interview. For instance, the senior author stressed in the latter setting that previous observations of these ratios were by no means widely accepted.

> 3 B
>
> *Spender:* . . . There's always criticism of one method. There are very few methods that are bomb-proof . . . What we did was another way of doing it . . .
>
> *Interviewer:* So there were people at that time who were casting doubt on P and Q's figures?
>
> *Spender:* You bet there were. And not merely on the *figures*, but on whether it happened at all. [Certain people] said, 'It just doesn't happen. There are no protons ejected.' [13–14]

In the introduction, no hint is given that the ratio mentioned in the text had been strongly criticised or the previous methods put in question. Whereas in the previous introduction the position of those opposed to the authors is briefly characterised in adverse terms, in this introduction it is entirely ignored. Our first author, Leman, presented his results as furnishing a test of and further support for a model already clearly superior to the previous, poorly worked out approach; even though informally he accepted other scientists' reservations about the central ideas of the model and their doubts about its empirical foundation as entirely reasonable. Similarly Spender depicted his results in the formal paper as an advance in method; even though he says that he was well aware that many scientists doubted whether previous observations were correct and even whether the phenomena which he was supposed to be measuring actually existed.

It thus seems that, not only does the characterisation of action and belief vary from one scientist to another, but also that these scientists offer quite different versions of action and belief in formal papers as compared to informal interviews. In both these papers, the scientific conclusions being proposed are made to appear as if they followed unproblematically from empirical evidence produced by means of impersonal experimental procedures. In informal talk in the interview, however, the scientific views of the authors and their actions as scientists are sometimes allowed to appear much more personal, open to debate and generally contingent.

What is left out of the formal account

A style is adopted in formal research papers which tends to make the author's personal involvement less visible; and the existence of opposing scientific perspectives tends to be either ignored or depicted in a way which emphasises their inadequacy, when measured against the 'purely factual' character of the author's results. As a consequence, the findings begin to take on an appearance of objectivity which is significantly different from their more contingent character in informal accounting. This formal appearance is strengthened by the suppression of references to the dependence of experimental observation on theoretical speculation, the degree to which experimenters are committed to specific theoretical positions, and the influence of social relationships on scientists' actions and beliefs, all of which are mentioned frequently in informal accounting.

Consider the following statements made by Leman during his interview. In these two quotations he is describing his reaction to the idea embodied in the model mentioned in introduction 1, when it was first suggested to him by the head of his laboratory.

> 3C
>
> He came running into the seminar, pulled me out along with one of his other post-docs and took us to the back of the room and explained this idea that he had . . . He was very excited. He was really high. He said, 'What if I told you that it didn't take any energy to make ATP at the catalytic site, it took energy to kick it off the catalytic site?' It took him about 30 seconds. But I was particularly predisposed to this idea. Everything I'd been thinking, 12, 14, 16 different pieces of information in the literature that could not be explained, and then all of a sudden the simple explanation became clear . . . And so we sat down and designed some experiments to prove, test this. [8]
>
> 3D
>
> It took him about 30 seconds to sell it to me. It really was like a bolt. I felt, 'Oh my God, this must be right! Look at all the things it explains.' [14]

In the formal paper we are told that experimental results suggested a model which seemed an improvement on previous assumptions and which was, accordingly, put to the test. In the interview, however, we hear of a dramatic revelation of the central idea of the model, which was immediately seen to be right, which revealed existing data in a new light and which led to the design of entirely new experiments. The author mentions in the interview that this major scientific intuition came to the head of the laboratory when he was 'looking over some old data'. But the essential step which so excited those concerned is depicted as conceptual and speculative rather than empirical and controlled. It was the act of perceiving new meanings in data which were already familiar. Furthermore, when the authors of the introduction refer to 'results which *suggest* a new model', they cite precisely those results which were, according to the informal account, actually produced *as a result* of the intuitive formulation of the central idea of the model. It appears, then, that the actions involved in formulating the model are characterised quite differently in the formal and the informal accounts. Whereas in the former the model is presented as if it followed impersonally from experimental findings, in the latter the sequence is reversed and the importance of intuitive insights is emphasised.

The formal and informal accounts also differ in their treatment of the author's degree of commitment to the model. No explicit reference is made in introduction I to the author's prior involvement in the model's formulation. The impression is given that the author is engaged in subjecting the model to a detached, critical test. Informally, however, significantly different statements were made.

> 3E
> When I arrived here, I thought that the clearest way of demonstrating that energy input served to promote ATP dissociation from the enzyme rather than the formation of a covalent bond, would be to show a change of binding affinity for the ATP upon energisation. Everything up to that point had been kinetic evidence ... and I felt some nice good thermodynamic data would help. [31]

> 3F
> It is a kind of shocking idea. 'Hey, everybody has been taught it takes energy to make ATP and now you are going out preaching it doesn't take energy to make ATP, it takes energy to get it off the catalytic site.' It was hard to sell ... I personally think that it's not proven, but I think it's pretty close. [19]

In quotation 3E, Leman does not describe himself as testing the model or trying to disprove it. Rather he portrays his actions in terms of looking for new kinds of evidence to furnish additional support. Similarly, in quotation 3F he stresses that, although only the smallest degree of uncertainty existed in his own mind, it was difficult to convince other scientists, who had not shared that initial revelation, that this model was required by the available evidence. Several phrases used by the author in the interview, varying from 'Oh my God, it must be right!' and 'it's pretty close to being proven' to 'some nice good thermodynamic data would help to demonstrate it', express strong commitment to the model. Yet in the formal account, he does not refer to his own involvement in the formulation of the model and implies a considerable degree of critical detachment: 'A critical test of this proposal would be to measure . . .'

The last phrase in quotation 3C, to the effect that he designed experiments 'to prove, test', the model, suggests that, informally, the interviewee did not distinguish clearly between testing and proving the model. The following additional passages illustrate how, in informal talk, he tended to approach the issue of testing theories. In quotation 3G, he is referring to his own work following on from that presented in the paper under discussion here. In quotation 3H, he is talking about the response of one of his opponents to critical tests of *his* theory.

> 3G
>
> *Interviewer:* Do you see your current work as testing out the alternating site model or filling in details?
> *Leman:* I think, well no. We *may* come up with additional data for the alternating site. But basically the aim is just to learn something about the catalytic site and not to test this further. [35–6]

> 3H
>
> He's tenacious . . . He's trying to accommodate data that doesn't agree with it by constructing a fairly complicated explanation. I think eventually he's got to give it up because I think it is probably wrong. What he does is, and this is not a bad type of technique, I'm not criticising him for it, when he hears something that doesn't agree with his ideas he tries to find an explanation. The problem is that he's constructed such a complicated explanation for this, that the whole thing should be dismantled and he should start again. [24]

We can see from these passages that, informally, the speaker did not insist that experimental work has to put the researcher's theoretical framework to the test and that he was willing to accept as legitimate what he saw as a dogged, *ad hoc* defence of a false theory. In view of his uncertainty as to whether the work reported in his paper was or was not devised to test the model, and in view of his acquiescence in quite different versions of research strategy, the reference in the introduction to carrying out a critical test can be seen to be only one possible description from several. In a context other than that of the research paper, it would have been quite appropriate to have characterised the same actions quite differently; for example, as an attempt to prove an interpretative speculation to which the author was strongly committed.

We have seen that Leman, in informal discussion, treats his adoption of this particular model of ATP synthesis as being brought about by his experiences in a specific laboratory and by his close contact with a particular group of colleagues. This recognition of how social relationships may influence the course of individual scientists' research is excluded from the formal paper. Thus introduction I refers simply to the fact that 'several laboratories' had produced results which supported 'the model'. Only by consulting the references could the reader observe that no more than two laboratories seem to have been involved and that the author's presence as a co-author of one of the cited papers show him to have been a member of one of these laboratories. Yet in informal discussion, the author stressed how significant for his research career was the period in that laboratory and how he has retained strong social links with its members and with their research.

> **3 J**
>
> I went to Bradshaw's lab. He had a very profound influence on me. That was really where I was educated. [4–5]
>
> **3 K**
>
> We were struggling with it. My students and I had all these diagrams all over and I wonder what would have happened if we hadn't gotten something in the mail. I wonder if we would ever have stumbled on it ourselves; probably not. But I got a preprint from Bradshaw; the Bradshaw, Willow and Stein paper, in which he proposed two cooperative catalytic sites and as soon as I saw it I liked it . . . This hadn't been published and we had the advantage of knowing it before it came out. Bradshaw was very kind and kept us up with what he was doing. [33–4]

This intellectual indebtedness and informal contact, which form natural topics for discussion in the interview, are not revealed in the introduction. The final paragraph of the introduction simply states that the empirical results which are to follow are difficult to assimilate by means of traditional assumptions and that they are more easily interpreted in terms of the model proposed by Bradshaw *et al.* A 'literal reading' of that final paragraph might be along the following lines: that the authors carefully examined the compatibility of their results with the major theoretical positions available in the literature and were led to conclude that one theory was shown to be clearly superior to the others in the course of an impartial appraisal. However, the account given informally is dramatically different. In the first place, the speaker says that he had already decided whilst in Bradshaw's laboratory that the traditional view was inadequate and that he never seriously considered interpreting his results within its frame of reference. Secondly, he reports that his experimental design was based on Bradshaw's original model, which he himself had played a part in formulating. Thirdly, he stresses that his acceptance in this paper of the revised version of the model proposed by Bradshaw *et al.* would have been impossible without fairly direct informal contact with Bradshaw.

Once again, it is clear that certain elements which are prominent in the informal account are left out of the formal introduction. There are, of course, bound to be differences of some kind between the informal and the formal accounts, if only because the former can be detailed and discursive whereas the latter are required to be brief and concise. Consequently, simply to show that differences exist does not itself take us very far. However, we have begun to show that these differences are not random, but systematic and meaningful. Certain clearly identifiable ways of characterising social action, which are treated as normal in ordinary discourse and in interviews, are consistently omitted from formal accounts; whilst other, opposite, attributes are emphasised. The underlying rationale whereby informal accounts are transformed in the course of formal accounting will become clearer in the next section.

Social accounting in methods sections

The existence of social accounting in experimental papers is most obvious in the sections on 'methods and materials'; for these sections consist mainly of highly conventionalised accounts of what the authors did in their laboratories in the course of producing their empirical results. The following quotations reproduce parts of the methods sections of the two papers discussed above. These sections are typical of the great majority of experimental papers in this area of biochemistry.

> **METHODS SECTION I (main author Leman)**
> Heavy beef heart mitochondria were prepared by the method of Wong and stored in liquid nitrogen. Well coupled mitochondrial particles were prepared by a modification of the procedure of Madden. These particles were used to prepare inhibitor-protein-depleted particles by centrifuging under energised conditions according to the method of Gale . . .
> In order to establish that ADP formation is the only rate-limiting step in our spectrophotometric assay for ATP hydrolysis, the following test was performed for each preparation of assay medium. Hexokinase and glucose were added to give a rate of absorbance change equal to or greater than that of the fastest ATP hydrolysis activity to be measured. The amount of hexokinase was then doubled, and the assay medium was considered adequate if the rate of absorbance change doubled . . .
>
> **METHODS SECTION II (main author Spender)**
> [A particular strain of yeast] was grown in continuous culture under conditions of glycerol limitation (Conran and Spender) or sulphate limitation (Hill and Spender). A variant of this yeast that does not require copper and has a cyanide insensitive terminal oxidase (Jason and Spender) was grown in continuous culture in a copper extracted medium . . . Harvested and washed cells (Conran and Spender) were converted

into protoplasts and mitochondria isolated as described by Castle. Protein was determined by the method of Sheridan. Measurements of respiration-driven proton translocation were made with the apparatus described by Mason and Spender in 1.0 ml of anaerobic 0.6 M-mannitol ... Polarographic measurements of P/O ratios were performed as described by Shoesmith, by using the experimental conditions of Spender ...

One of the most noticeable features of these passages is the way in which the specific actions of the researchers in their laboratories are expressed in terms of general formulae. Constant reference is made to methodological rules formulated by other scientists; and many of the authors' actions are not described at all, but are simply depicted as instances of these abstract formulae. This is sometimes so, even when the authors recognise that they have actually departed from the original formula. Thus not only do we find 'mitochondria were prepared by the method of Wong', but also 'particles were prepared by a *modification* of the procedure of Madden'. Where the authors' laboratory procedures are seen as introducing new practices for which existing rules do not provide, these practices are themselves presented as rule-like formulations as in the second paragraph of methods section I, which can then be used by other scientists in the course of *their* work. Informally, the principle which was usually said to guide authors in writing methods sections was that they should provide enough information for other scientists to repeat the authors' relevant actions and get the same results. As Spender stated: 'In the *Biochemical Journal* they have a separate section and you have to give sufficient detail there to enable any competent scientist to reproduce your experiment.' Thus the form of accounting used to depict scientists' actions in methods sections seems to be more or less explicitly an attempt to extract certain invariant dimensions from the unique, specific actions carried out by particular researchers in particular laboratories and to embody these dimensions of action in general, impersonal rules which can be followed by any competent researcher.

Methods sections, then, appear to be formally constructed as if all the actions of researchers relevant to their results can be expressed as impersonal rules; as if the individual characteristics of researchers have no bearing on the production of results; as if the application of these rules to particular actions is unproblematic; and as if, therefore, the reproduction of equivalent observations can be easily obtained by any competent scientist through compliance with the rules. In the course of informal talk, however, each of these notions is repeatedly contradicted. For instance, it is frequently noted that exact compliance with another's methods and exact replication of their results is virtually impossible.

3 L

When you write the paper which says how you did it, the ground rules are that you write it in such a way that other laboratories could reproduce your work and your conditions. Now that of course is impossible. There are all sorts of things that you don't know about, like 'finger factor', the local water, built-in skills, which you have taken for granted. But you try and do it anyway. [Spender, 16]

Or as another respondent put it:

3 M

Ideally, the scientific paper should make it possible, assuming that a library is available, for a Martian to come and do your experiment. But that's largely wishful thinking. [Richardson, 17]

Methods sections give the impression that the application of methodological procedures is a highly routinised activity, with little room for individual initiative and variability. Informally, however, scientists stressed that carrying out experiments is a practical activity requiring craft skills, subtle judgements and intuitive understanding. They talked of particular researchers having 'good hands' or 'a feel' for laboratory work.

3 N

You get a feel for what you need. I can tell you a story about this. I went to the workshop once to get something made. There was no way they could do anything for me for a week or a month. They were making something for Dr X. I said 'What are you making for Dr X?' 'Dr X requires his water bath to operate at 36.5°C and *nothing else*.' And they were having a hard time actually. I said, 'That's ridiculous.' And I consulted with Dr X and he produced this paper showing that in this experimental protocol, they'd worked at 36.5°C. It didn't matter a damn really, whether it was 35 or 40°C, as long as you stayed roughly where you were. Dr X was not an experimenter and no longer does any. If you are an experimenter you know what is important and what is not important. [Spender, 24]

When discussing laboratory practice informally, authors emphasised that dependence on an intuitive feel for research was unavoidable owing to the practical character of the actions involved. Such actions cannot be properly written down and they can only be understood satisfactorily through close personal contact with someone who is already proficient.

3 P

How could you write it up? It would be like trying to write a description of how to beat an egg. Or like trying to read a book on how to ski. You'd just get the wrong idea altogether. You've got to go and watch it, see it, do it. There's no substitute for it. These are *practical* skills. We all know that practical skills are not well taught by bits of paper. Could you write a dissertation on how to dig your garden with a fork? Far better to show somebody how to stick the fork in and put your boot on it. [Spender, 26]

In addition, scientists pointed out that many aspects of laboratory practice are traditional; in the sense that they are done because they are customary and are assumed, without detailed analysis, to be adequate for the task in hand.

> **3Q**
>
> *Interviewer:* One of the things we find difficult in reading those papers is understanding just *why* you have done certain things or used certain chemicals. Is there a convention about that?
>
> *Spender:* The convention is that you normally use what you used last time round. You don't want to change. Let's take an example. We want to suspend mitochondria in some medium. Now if you put mitochondria in water they swell and burst. So they need support. Why did we use 0.6 ml? 0.6 ml is about right. Why did we use mannitol and not sucrose or something else? Well, because somebody in Japan 10 years ago had published the first paper on making mitochondria and he used mannitol. I don't know why they used mannitol. They may have been given it. Or they may have found it was better. Or maybe it was what they started with and they didn't want to change. So that's why we used mannitol. We saw no good reason to change from the original recipe. And 'recipe' is the right word. It's like cooking. [Spender, 18]

As a result of their emphasis on the role of customary practice and on learning by example, it is not surprising that many authors said that it is often extremely difficult to specify in full the actions relevant to the production of their results. Even when it is possible for a scientist to work out from the formal paper what, for practical purposes, counts as a repetition of another's methods, unless he is working on something very similar it is likely to 'take him an awful long time. Because there are so many mistakes you can probably make, I suspect. And I wouldn't even know what they are, you see, that's the snag. They'd be things I'd take for granted.' [Spender, 20]

Scientists who belong to the same research network often claim, of course, to have succeeded in translating the content of their colleagues' formal methods sections into effective laboratory practice. But they also acknowledge that when these formal accounts pass outside the small, specialist community to scientists who do not share the same background of technical assumptions and who have not experienced close personal contact with its members, this process of translation becomes much more difficult.

> **3R**
>
> One is telling the general reader very roughly how the experiments are

done and the specific reader, that is anyone else who is working in the same field, which of the things he already knows about you have chosen to do. I mean, it would not enable an intelligent scientist in another field to set about doing those experiments . . . [Richardson, 20]

3S

From my own experience of trying to read back old papers, it can be a nightmare sometimes, trying to work out what they actually did . . . People outside the field don't even know where to get the reagents from. If I look at a paper in molecular biology, where they're using all these fantastic antibiotics, I wouldn't know where to start unless they listed the sources of supply . . . [Spender, 16]

It is clear, then, that the accounts of scientists' actions which appear in the methods sections of research papers differ radically from the accounts of the same actions offered informally. Whereas formal methods sections contain highly abstract versions of scientists' research activities in the form of impersonal rules, with no attempt to specify how these rules are interpreted in practice in particular instances, scientists' informal accounts emphasise that these rules depend for their practical meaning on the variable craft skills, intuitions, customary knowledge, social experience and technical equipment available to individual experimenters. It is also clear that scientists themselves are able to describe some of the differences between their formal and informal accounts of laboratory practice. They recognise that what they regard informally as crucial aspects of their actions are omitted from the versions given in research papers. They stress that the ostensible objective of methods sections is unattainable and they imply that the meanings given to the rule-like formulations employed in methods sections vary in accordance with readers' membership of specific social groupings.

Empiricist and contingent repertoires

We have shown that, in the material examined above, two significantly different forms of social accounting are available to scientists and are selectively employed by them. The formal research literature is dominated, we suggest, by an empiricist repertoire. Although this repertoire is also used widely, as we shall see, in situations like informal interviews and ordinary conversations, on such occasions it is frequently supplemented by forms of contingent discourse not normally found in the research literature.

Our reference to the existence of an empiricist repertoire is based on the observation that the texts of experimental papers display certain recurrent stylistic, grammatical and lexical features which appear to be coherently related. As we have seen, in research papers experimental data tend to be given chronological as well as logical priority. Neither the author's own involvement with or commitment to a particular analytical position nor his social ties with those whose work he favours are mentioned. Laboratory work is characterised in a highly conventionalised manner, as instances of impersonal, procedural routines which are generally applicable and universally effective. Although the content of experimental papers clearly depends on the experimenters' actions and judgements, such papers are overwhelmingly written in an impersonal style, with overt references to the authors' actions and judgements kept to a minimum. By adopting these kinds of linguistic features, authors construct texts in which the physical world seems regularly to speak, and sometimes to act, for itself. Empiricist discourse is organised in a manner which denies its character as an interpretative product and which denies that its author's actions are relevant to its content.

When the author is allowed to appear in formal texts, he is presented either as being forced to undertake experiments, to reach theoretical conclusions, and so on, by the unequivocal demands of the natural phenomena which he is studying or as being rigidly constrained by invariant rules of experimental procedure which are, in turn, required by the nature of the physical world. Each scientist's actions and beliefs, no matter how inconsistent they appear to be with those of other researchers, are presented as those of any competent scientist. The guiding principle of this repertoire appears to be that speakers depict their actions and beliefs as a neutral medium through which empirical phenomena make themselves evident. The stylistic, grammatical and lexical resources of the empiricist repertoire can be seen as related to this guiding principle in the sense that they are necessary features of texts which are consistently depicting participants' professional actions and scientific views as inevitable, given the realities of the natural world under study. We call this repertoire the 'empiricist repertoire' because it portrays scientists' actions and beliefs as following unproblematically and inescapably from the empirical characteristics of an impersonal natural world.

The selectivity of this kind of representation becomes evident when we compare scientists' formal accounts with the characterisations of the same acts given by the same scientists as they engage in informal discourse. As we have seen, at certain points in their interviews, scientists presented their actions and beliefs as heavily dependent on speculative insights, prior intellectual commitments, personal characteristics, indescribable skills, social ties and group membership. Not only was the general style of participants' informal discourse much more personal and idiosyncratic, but in certain passages they used the wider range of stylistic, grammatical and lexical resources to be found in informal talk to construct accounts of their own and others' actions and beliefs that were radically different in content from those appearing in comparable formal texts. In particular, the wider range of interpretative resources employed in informal talk allowed scientists to construct accounts in which the connection between their actions and beliefs and the realm of biochemical phenomena appeared much less direct and much more dependent on other variable influences.

Thus scientists' informal talk about action and belief was often much more contingent, in the sense that speakers gave accounts in which it was accepted that their professional actions and scientific views could have been otherwise if their personal or social circumstances had been different. We refer to this form of discourse as the contingent repertoire. Its guiding principle is in direct opposition to that of the empiricist repertoire in that it enables speakers to depict professional actions and beliefs as being significantly influenced by variable factors outside the realm of empirical biochemical phenomena. When this repertoire is employed, scientists' actions are no longer depicted as generic responses to the realities of the natural world, but as the activities and judgements of specific individuals acting on the basis of their personal inclinations and particular social positions.

The identification of these interpretative repertoires is a first step in making sense of the ordered variability of scientific discourse. It helps us to begin to understand how scientists, as they reproduce different kinds of context within the social world of science through the use of different linguistic registers, come to generate discrepant versions of action and belief. At this point, however, before we take the analysis further, we need briefly to reflect on the nature of the context within which our respondents' informal talk has been generated. We have to consider whether the discourse of sociological interviews is likely to be special in some way and whether we can properly claim to be comparing scientists' formal with their informal discourse. Clearly, it follows from our general analytical position that interview talk will differ to some degree from that occurring on other kinds of occasion. It has to be accepted in principle, therefore, that the contingent repertoire could be a resource on which scientists draw only when they are engaged in interviews or that it is used only during discussion between scientists and laymen. If the contingent repertoire appeared only in these contexts, the analysis would have limited significance. However, as we shall see more clearly in later chapters, this seems in fact not to be the case.

The contingent repertoire and the distinction between the two repertoires appear to play a major role in various kinds of naturally occurring informal discourse among scientists. It is clear, for example, that scientists employ both the repertoires identified above in a similar manner in conference discussions as well as in interviews. Moreover, we will see below that certain kinds of scientific humour trade upon participants' informal use among themselves of the two repertoires. Thus, although the forms of discourse appearing in interviews and in 'naturally occurring' situations undoubtedly differ, these differences seem irrelevant to the broad observations we have made so far. We will accordingly, for the moment, treat interview discourse as typical of a loosely defined and wide-ranging context of informal interaction involving scientists. Nevertheless, we emphasise that this is no more than a preliminary definition. Undoubtedly interview talk will be found to differ systematically in certain as yet unknown respects from other forms of scientific discourse. Eventually, therefore, particularly as more direct recordings of naturally occurring talk among scientists become available, it should be possible to refine our concepts and to begin to specify which facets of interview talk are peculiar to interviews and which are generic to informal discourse in science.

Scientists' accounts of interpretative variation

If there are systematic variations in scientists' discourse, of the kind we have suggested, it would be surprising if participants were entirely unaware of and unable to comment upon them. Our respondents were, in fact, not only able to describe some of the contextual differences in discourse which we have illustrated, but they were also able to furnish what they presented as adequate explanations of these differences. These explanatory accounts themselves exemplify how scientists draw flexibly on both the empiricist and the contingent repertoires in the course of informal talk. We can, therefore, extend our analysis by examining these accounts. In the following passage, Leman is explaining why the scientific literature omits to mention the author's personal involvement. We have numbered each statement in this and the following quotation for ease of reference.

> 3T
> 1 Everybody wants to put things in the third person. So they just say, 'it was found that'. 2 If it's later shown that it was wrong, don't accept any responsibility. 3 '*It* was found. I didn't say I *believed* it. *It* was found.' 4 So you sort of get away from yourself that way and make it sound like these things just fall down into your lab notebook and you report them like a historian ... 5 Of course, everybody knows what's going on. 6 You're saying, 'I think'. 7 But when you go out on a limb, if you say 'it

was shown that' or 'it is concluded' instead of 'we conclude', it should be more objective. 8 It sounds like you are taking yourself out of the decision and that you're trying to give a fair, objective view and that you are not getting *personally* involved. 9 Personally, I'd like to see the first person come back. 10 I slip into it once in a while. 'We found.' 11 Even then I won't say 'I'. I'll say 'we' even if it's a one-person paper. 12 Can spread the blame if it's wrong [*laughs*]. [Leman, 57–8]

Attention is drawn in this passage to the use of impersonal formulations and the tendency to avoid revealing one's personal involvement. The reason the speaker gives for this stylistic convention is that it enables authors to avoid accepting any responsibility for errors (3T2). Yet he appears almost immediately to contradict this explanation, when he suggests that nobody is ever actually misled into thinking that no one was responsible for a paper's contents because it made little or no mention of human agents (3T5–6). Thus the rationale offered by this interviewee for this feature of formal texts seems to be internally inconsistent and unconvincing. The inconsistency seems to occur because the speaker combines an explanation of formal discourse in contingent terms (3T5–6,11–12) with an interpretation which stays close to the guiding principle of the empiricist repertoire (3T4,8).

Our second author also emphasises the impersonal character of research papers. But he argues that scientific papers have this characteristic because they are devised in terms of a mythical conception of scientific rationality.

3U

Spender: 1 I think the formal paper gets dehumanised and sanitised and packaged, and becomes a bit uninteresting. 2 In some ways I like the old ones, where a chap says, 'I did this and it blew up in my face' . . . 3 Some of the charm has certainly gone.
Interviewer: 4 Why do you think that is?
Spender: 5 One is a myth, that we inflict on the public, that science is rational and logical. 6 It's appalling really, its taught all the way in school, the notion that you make all these observations in a Darwinian sense. 7 That's just rubbish, this 'detached observation'. 'What do you *see*?' 8 Well, what *do* you see? God knows, you see everything. 9 And, in fact, you see what you want to see, for the most part. 10 Or you see the choices between one or two rather narrow alternatives. 11 That doesn't get admitted into the scientific literature. 12 In fact, we write history all the time, a sort of hindsight. 13 The order in which experiments are done. All manner of nonsense. 14 So the personal side does get taken out of this sort of paper. 15 Maybe it's felt that this isn't the place for it to be put. I don't know . . . 16 Sometimes you get more of the personal side in reviews. 17 Some of them are quite scandalous actually, once you can read between the lines.

Interviewer: 18 Do you think there would be any disadvantages in allowing that sort of thing back into the formal literature?
Spender: 19 I don't know. It depends what the purpose of the literature is. 20 If the purpose of the literature is to describe what you did, why in scientific terms you did it – I mean, not because you want to do some bloke down or you want to advance your own career or get a quick paper out just because there's a grant application coming up soon. 21 All these are valid reasons, but they're never admitted to. 22 If the publishing reason is to present the science, what you did and what the conclusions were, then there really isn't much room for the emotive side. 23 If I'm writing a paper [I don't say] 'I don't think that Bloggins understands electro-chemistry because he's a dum-dum.' 24 I might say 'This was overlooked by Bloggins *et al.*' 25 I won't say why I think they overlooked it. 26 I'm afraid it's *gone* and it's not going to reappear here. 27 Probably it shouldn't reappear. I guess it reappears in other places. 28 And we still know what's going on. 29 We just don't make it public. [Spender, 32–3]

Like the previous speaker, this scientist stresses informally that researchers can 'read between the lines'. For example, they will read Bloggins' paper bearing in mind that he is, in their opinion, a dum-dum. In our terms, participants are being represented as translating the formal text into their more extended informal repertoire. Initially, Spender's claim seems to be that when scientists construct research papers they reinterpret and re-order their prior actions so as to make them appear to fit an empiricist myth of scientific action. He seems to suggest that this myth has been devised for public consumption (3U5–6,29). Spender's description of the observable features of formal discourse, like that of Leman, is entirely consistent with our own (3U6–7,11–14, 23–5). Furthermore, Spender begins to formulate, in a very preliminary fashion (3U7–14), an alternative to the empiricist conception of scientific action and belief, which is presented as a more accurate portrayal of what actually happens in science and which is couched in terms of the contingent repertoire. But he fails to take this line of thought very far beyond the basic assertion that science in practice differs considerably from the conception embodied in research papers and that in practice scientists are greatly influenced by personal factors. Moreover, in the second half of the passage he clearly returns to the more traditional, empiricist view of science (3U20–2). When he says that 'if you publish in order to present the science, there isn't much room for the emotive side', the speaker seems to be ignoring what he had just said about scientists 'seeing what they want to see' and constructing acceptable, non-emotive versions of their actions after the event. Thus, like the previous respondent, Spender fails to produce a coherent, overall account of the differences between formal and informal discourse. Both speakers generate inconsistencies as they move somewhat erratically between empiricist and contingent versions.

We suggest that these interpretative difficulties arise because the speakers in these passages present interpretations of the formal realm in contingent terms along with interpretations couched in its own empiricist terms. Because both our respondents had been giving extended contingent accounts of their research practices immediately before being asked to reflect on the character of formal discourse, it seems reasonable to suppose that it was difficult for them to avoid mentioning the 'conventional and unrealistic' features of that discourse. In other words, both respondents are inclined by the discursive situation created during the interview to furnish contingent accounts of the actions supposedly lying beyond and behind formal scientific discourse. As a result, as we can see in the passages above, the characteristics of empiricist discourse are to some extent made to appear merely rhetorical. Yet, if our analysis is broadly correct, the empiricist repertoire is such an important interpretative resource, in informal as well as formal settings, that its rejection is likely to generate interpretative problems in any subsequent talk about science. Moreover, as we will see in the next chapter, speakers continually construct empiricist accounts of their own scientific position. We would, therefore, expect scientists to draw quickly back from or to qualify any talk in which they can be heard as generally undermining empiricist versions of scientific action. Thus we can understand the interpretative inconsistencies in our respondents' explanations of the nature of formal discourse as following from their socially generated (i.e. discursively generated) use of two formally incompatible interpretative repertoires to provide accounts of action.

The passages contained in this section begin to reveal some of the interpretative complexity and variability of informal scientific discourse, in so far as interviews can be taken to represent such discourse. This apparent variety of informal discourse poses an important issue for our analysis. For, although it is clear that the restricted and highly conventional formal discourse of science does display evident interpretative regularities, it is by no means so obvious that informal discourse is systematically organised to anything like the same extent. On the basis of our analysis so far, it could be argued that what we have termed the 'contingent repertoire' is merely a residual category, containing a *melange* of disparate elements which have in common only the fact that they do not appear in the formal literature and are sometimes difficult to reconcile with its empiricist vocabulary. Our next step, therefore, must be to begin to look in more detail at scientists' informal discourse, in order to observe whether it displays recurrent interpretative forms and to explore whether scientists employ the contingent repertoire as a coherent discursive resource. In short, we will now start to move from a fairly crude identification of broad interpretative repertoires to a more fine-grained examination of the orderly production of social meaning in science.

4

••

Accounting for error

The biochemists we interviewed, in talking about their research, devoted much effort to distinguishing between scientific truth and error in their field, and to explaining why particular scientists had adopted correct or incorrect theoretical positions. Their talk also contained 'good reasons' why they should pay particular attention to truth and error. For instance, they pointed out that when there are two or more competing 'theories' available in a given area of investigation, each will usually lead to the design of quite distinctive experiments. Accordingly, if a scientist is to do satisfactory experimental work, and all our respondents had published experimental results, it is of crucial importance that he chooses the theory which is most nearly correct. Similarly, in devising his experiments, he must decide on the adequacy of others' observational claims, because acceptance of some claims rather than others will have a direct bearing on what experiments should be undertaken next and what results one should expect to achieve.

During interview talk, then, these scientists regularly identified each others' scientific positions and took note of each others' theoretical and experimental errors. In addition, they often tried to account for their colleagues' errors, that is, they provided versions of participants' actions and beliefs which made these errors readily understandable. In this chapter, we will focus on our respondents' attempts to account for theoretical errors. It is probable that retrospective accounts, including accounts of error, occur more frequently in interviews than in 'ordinary' informal talk among scientists. Nevertheless, it is clear from studies of numerous types of conversational data that the orderly reconstruction of past action and belief, exemplified in this chapter by scientists' accounts of error, is a recurrent feature of ordinary talk. Silverman, in particular, has drawn attention to the similarity between interviews and informal conversations in this respect.

> A final feature of interviews that I want to address arises in a common characteristic of talk: in their accounting activities members concern themselves with displaying what will currently be understood as rational grounds for past actions and as rational explanations of past social

scenes, i.e. they seek to display their purported 'sensible' and 'reasonable' character. Furthermore, this sensible character is found in what 'finally' is seen to transpire – so that, for all practical purposes, *the meaning of the past is found in the present.* [Emphasis added][1]

The present outcome of past action in relation to which scientists tend to organise their rational reconstructions is overwhelmingly the scientific correctness of the speaker's own current intellectual position. This feature is so widespread in our data that we propose it tentatively as the fundamental principle of social accounting in science. This does not mean, however, that each scientist's present intellectual position provides a fixed reference point for the construction of his discourse. For participants' scientific views, like their other interpretative resources, are continually reformulated in the course of ordinary talk and textual production. Thus the principle proposed above means that each scientist organizes his accounts of action and belief in ways, appropriate to the particular interpretative context, which explain, justify and validate the version of his scientific position furnished in a specific passage of talk or in a particular unit of discourse. We will see that in the present chapter this principle applies consistently in the case of scientists' accounts of error.

Our aim in the rest of this chapter is to document the recurrent interpretative features which appear in passages where scientists are making sense of theoretical error. We will try to identify the particular features of scientists' reconstructions whereby the occurrence of scientific error is made understandable. Although our data are primarily taken from interviews, we will also provide some indication that the same interpretative form is used by scientists in other kinds of discourse.

Some examples of accounting for error

In this section, we offer some instances of the kind of interpretative work by scientists with which we are concerned in this chapter. We ask the reader to examine them carefully and to reach his own preliminary conclusions about their interpretative structure before moving on to consider our analysis. After we have identified in the next section what we take to be the main features of such accounts, we will proceed to examine further examples in greater detail and to extend the analysis throughout the rest of the chapter.

4 A

I had no axe to grind. It's an advantage not being able to contribute in the theoretical sense. I mean, you don't feel that you have time and publications and reputation based on previous contributions and it's very easy to go the way the evidence seems to point. It leads to more flexibility.

People like Gowan and Fennell especially and Milner, certainly had many publications and they discussed one theory as they went along and they had a lot invested in that field and I think they were psychologically a little bit reluctant to follow the lead of – utterly new, strange and different coming from somebody else completely. Certainly that remains the case with Pugh. [Miller, 22]

4 B

This is the effect of removing membrane potential. Now we ask what happens if we *now* prevent, within *this* system, prevent a hydrogen ion accumulation inside, when we don't think we can have any membrane potential. Now you will have, you will have people, particularly people at [a particular university], who will give you absolute hell about those experiments. But the people at [that university] are wrong. The people at [that university] are wrong because they are too damned dogmatic. They think this is an insuperable barrier to the chemiosmotic theory or at least it is beyond the range that's acceptable to the chemiosmotic theory. And that's no way to do science. The facts are pretty clear experimentally and these people are sort of misquoting the fact. [Southgate, 20–1]

4 C

Fennell had become quite an influential person actually and he was Professor of Biochemistry and very much an anti-Spencer man. I always remember when I was a post-doc that Fennell came down to give a talk about why the Spencer scheme was wrong and it was just a load of nonsense, you know. It really was and I remember it made me so angry. I remember having violent arguments with him. Of course, Snow [the speaker's supervisor] couldn't understand my arguments at all and certainly Fennell didn't, because he was putting up such ridiculous things. I might say that now Fennell believes in the Spencer scheme. But he'd built his whole career up on opposing it and he believed in the chemical intermediate hypothesis . . . in the case of people like Fennell, who were forceful people in bioenergetics, they didn't really understand the . . . simple thermodynamics, really. They didn't really understand it and because they were forceful people the controversy built up. [Grant, 23–4]

4 D

That was another strength of the [chemiosmotic] theory. You could take somebody else's experiments and they could be entirely reinterpreted in a way which was more simple than the one offered by the authors . . . But Waters didn't believe any of it. None of it. He'd been brought up with the chemical theory. He'd made several contributions to that. He'd interpreted all his work on [a particular reagent] in terms of it, in a complicated way. He was a great friend of Watson's. He knew Gowan. It was America anyway. The chemiosmotic theory, as far as he was concerned, was a little bit of a joke. Perhaps an irritating one. And there was this damn Englishman interpreting everything. [Barton, 11]

4E

There is, to my way of thinking, not a single piece of evidence that will bear close examination for the Spencer model. And the *crucial* piece of evidence – Ditchburn has written in the last *TIBS* on the membrane potential and he has looked for that membrane potential for the last 15 years. There is no membrane potential, period. That's a source of embarrassment to everybody, because the Spencer model requires it. He's never seen it and they have pilloried him. Everybody is looking under the bed: did you do this? And did you do that? And he goes back each year and he does all the controls that they claim he should do, and he does them. And he still gets the same answer – it isn't there. Now they say, well – it's like religion. People don't know *why* they believe certain things. They believe them. Their fathers believed them. Their mothers believed them. So they believe them. Its purely irrational now. There's no-one I know can make a reasoned case for the Spencer model at the present time. [Pugh, 21]

4F

I do think it was the little grey – first let me preface this by saying I am a good friend of Spencer's . . . So I don't choose to say anything – I just try to give you the facts as I see them and I do think that [Spencer's] little grey books, never having to go through a review, were much more extensive and comprehensive than you could have got into the literature. And I do think that they stimulated a tremendous number, well a lot of students, who really pushed the hypothesis and created an aura of fact, when there might not have been fact. You know, the group at [a particular university], Richardson, Crosskey and Burridge, did a tremendous amount to promote the idea, without ever questioning the things that Milner or myself or Lucas might have questioned about – 'Is it right?' Instead, they took the stoichiometry, ratios of hydrogen ions to oxygen, as OK and they were not OK. And things like that. But I do think that a lot was based upon the fact that the grey books were so comprehensive and well written. Spencer writes well, no doubt about it . . . I do think that for ten years the very strong support was forthcoming without coming down to the *critical* issues. [Gowan, 2]

4G

I think that there was just a tendency for people to try to give the impression that they were right. And a lot of us found that they were betraying us, you know, that they were really being very dogmatic about their views and they had very strong personalities and they were wrong. I think that that's one of the things that I probably discovered at an early enough age to where I could reorient my whole way of approaching things and not worry about what these people were saying and in fact attack them every chance I got and really to try to cut them to pieces to make them get down to just how you can say such and such. Where is the data for this? How can you exclude this? And then you found out that some of them had hearing problems. Perry could never hear what I had to say. He always had a hearing problem every time I asked him a question at the meetings. [Carless, 27]

4H

We usually use enzyme that's been depleted before we make the measurements. There are lots of things you have to take into account and there are very strong individuals in the field who want to interpret everything in terms of their theories. Of course, those are the other guys, not us. We're interpreting it even, balanced [*general laughter*]. The other ones are the ones who are doing that. When you try and bend the data like that sometimes you don't take into account everything, too. Its complicated. There are lots of unknown factors still to be discovered. [Hargreaves, 51]

The asymmetrical structure of accounts of error

During each of these passages the speaker (a) identifies the views of one or more scientists as mistaken and (b) provides some kind of account which enables us to understand why the scientist(s) adopted an incorrect theory or failed to accept a correct theory. Any passage which displays these two features is an example of 'accounting for error'. In all the passages above, the speaker's own view is taken to be synonymous with the correct scientific view. However, different speakers endorse different, and sometimes apparently diametrically opposed, positions. The first six quotations focus fairly clearly upon specific theories, in that they refer explicitly to the chemiosmotic or chemical intermediate hypotheses, to Spencerian or anti-Spencerian views, or to scientists who are frequently cited as advocates of specific theoretical positions. If we identify speakers' positions in these passages in relation to the chemiosmotic hypothesis, it appears that 4A, 4C and 4D are pro-chemiosmosis, that 4E and 4F are anti-chemiosmosis, and that 4B is difficult to categorise. Although 4G and 4H are rather more general in character, they also distinguish unequivocally between the speaker's correct view or scientifically proper research strategy, on the one hand, and a loosely defined collection of false views, on the other hand. Thus all these passages involve a marked contrast between correct and incorrect views of the phenomena of oxidative phosphorylation.

Another feature of these accounts is that speakers link the correct view directly to experimental evidence. In the sense in which we have used the word 'empiricist' in the previous chapter, each respondent presents his own position in empiricist terms. Each speaker presents his theoretical position as an unmediated expression of the natural world, in so far as that world has revealed itself in the findings of controlled experiments. For example, Miller says that because he had no axe to grind, it was 'very easy to go the way the [empirical] evidence seems to point' (4A). Similarly, Southgate begins his passage with an emphatic statement of what is shown to be the case experimentally and he states later that 'the facts are pretty clear experimentally', even though other scientists are unable to recognise them (4B). Barton in quotation 4D, having previously reviewed the more obvious experimental basis for the chemiosmotic theory, goes on to maintain that that theory is actually confirmed by what other people have (mis)interpreted as counter-evidence (4D). Pugh and Gowan, although they display totally different theoretical commitments to those of the previous speakers, also base their theoretical contentions directly on a personal reading of the empirical evidence which they present as if it were unproblematic. Pugh does this more dramatically, claiming that there is not a shred of evidence in favour of chemiosmosis and that one of the major constituents of the theory does not exist: 'There is no membrane potential, period' (4E). Gowan is more restrained. Nevertheless, he fits the general pattern in organising his account as if he had privileged access to the empirical world. Thus, the chemiosmotic theory was based on 'an aura of fact, when there might not have been fact'. Similarly, the theoretical claims of the chemiosmotic theory about stoichiometries and other matters are treated as being simply incorrect (4F). Other speakers draw attention to the importance of basing theoretical claims on the data and the widespread failure on the part of other scientists to do this (4G); or to the intellectual confusions characteristic of those who did not see the realities of the natural world with the accurate perception of the speaker (4C).

Although these speakers in aggregate are advancing a wide variety of conflicting views about a fairly narrow range of biochemical phenomena, in these passages they all speak as if their own position is an unproblematic and unmediated re-presentation of the natural world. In contrast, the actions and judgements of those scientists who are depicted as being or as having been in error are characterised and explained in strongly contingent terms. Their false claims about the natural world are presented as being mediated through and as understandable in terms of various special attributes which they possess as individuals or as certain kinds of social actor. For instance, scientists are presented as being in error because they are 'strong individuals who want to interpret everything in terms of their theories' and who, consequently, 'bend the data' (4H). Alternatively, they are characterised as 'strong personalities' (4C,4G), 'dogmatic' (4B,4E) and inclined to avoid awkward questions (4G), as being misled by publications which had not been subject to proper refereeing (4F), as irrational (4E), or as having too much invested in a theory to give it up (4A,4C). Even something as superficially irrelevant as being in America can be cited as a reason why a particular scientist got it wrong (4D). As we have seen in the previous chapter, the depiction of a scientist's actions in contingent terms does not in itself prevent those actions from appearing scientifically proper or from being associated with correct belief. Scientists do not necessarily undermine their accounts of laboratory practice, for example, by couching them in this manner. In accounts of error, however, the contingent representation of scientists' actions and beliefs is organised in such a way that it effectively removes the beliefs in question from the realm of scientific legitimacy. As Southgate puts it: 'that's no way to do science'. In other words, accounts of error are typically organised in a manner which not only explains scientific error by linking it to various 'non-experimental' factors, but in so doing explains it away.

In the passages above, these references to contingent factors are presented as if they explain, even though they do not spell out in detail, how it is that other scientists reached wrong conclusions. This is done through the employment of both the empiricist and the contingent repertoire within accounts which have an asymmetric structure; that is, the speaker's own empiricist speech is given interpretative precedence and provides an unquestioned context in relation to which other scientists' claims are to be classified, explained and repudiated. The speaker's presentation of his own views as identical with the discernible realities of the natural world furnishes the only viable, properly scientific frame of reference, in relation to which others' divergent views have to be taken as clearly false and in need of explication. To put this another way, each speaker who formulates his own position in empiricist terms, when accounting for error, sets up the following interpretative problem: 'If the natural world speaks so clearly through the respondent in question, how is it that some other scientists come to represent that world inaccurately? What is it about such speakers which prevents the natural world from representing itself properly in *their* speech?' This implicit question is resolved in accounts of error by the assertion that the views of these other scientists are being distorted by the intrusion of non-scientific, that is, non-experimental, influences into the research domain. The lexicon of the contingent repertoire is used to identify non-experimental factors which are probably mentioned regularly in the ordinary small talk of science and which can account plausibly for deviations from scientific accuracy. Thus the introduction of the contingent repertoire resolves the speaker's interpretative dilemma by showing that the speech of those in error, although it is not fully scientific, is easily understood in view of 'what we all know about' the typical limitations of scientists as fallible human beings. In accounts of error, then, the empiricist versions of correct belief provide instructions for the interpretation of contingent elements. Because contingent factors are mentioned only in the case of false belief, because they are directly contrasted with the purely experimental basis of the speaker's views and because their power to generate and maintain false belief is taken as self-evident, the contingency of scientists' actions and beliefs is made to appear anomalous and as a necessary source of, as well as an explanation of, theoretical error.

So far in this chapter, we have tried to give the reader an opportunity to scrutinise some of our data for himself and then to indicate in broad terms the kind of recurrent interpretative structure which we suggest is evident in that material and in the passages to be presented below. We will now explore some specific passages in more detail and begin to provide a firmer basis for our analysis of accounting for error. We should perhaps make it quite clear that terms like 'correct belief' and 'error' are intended to convey our understanding of particular respondents' statements, as expressed in interview transcripts, letters and papers, about the validity of their own and other biochemists' scientific views. They do not refer to *our* assessments of the biochemical knowledge-claims under discussion by participants.

The flexibility of accounting

A regular pattern in our biochemists' accounts of error, which we have already observed above (4B, 4C and 4F), is that the speaker contrasts his own experimentally based scientific appreciation with other researchers' 'failure to understand the issues' and then goes on to explain this failure by referring to various social and/or psychological characteristics of those concerned. The following quotation, from a relatively young researcher who described himself as having favoured Spencer's chemiosmotic ideas since he first entered the field, provides another example.

> **4J**
> I was just one of a number of people who were working with these new ideas. It just seemed that everything that we did could be explained satisfactorily by Spencer's theory . . . So we said, if this idea is right then we ought to be able to show such and such a thing, and we would go ahead and do it and it would work . . . Maybe in some ways we were a little bit dogmatic. I found it very interesting because, as a Ph.D. student, I was meeting guys like Gowan and I was able to say to them. 'No, you've got this wrong. This can be explained much more easily in this manner.' I found that people like Gowan didn't really understand what was going on, in terms of this hypothesis. We were just able to explain a lot of things

and do a lot of things, all in terms of Spencer's theory. It was only later that we started to get a little bit more critical of it . . . My impression was that there was certainly a lot of prejudice involved. Gowan is a good example because he was at the forefront in those days, a very important man. He'd done a lot of good work in the 1950s and he'd got his own models of energy coupling. I think he was probably quite defensive about those ideas. So he was reluctant to accept the chemiosmotic hypothesis in the first place. But not only that, I think he was also reluctant to put effort into understanding the details of it. It was fairly complicated . . . Gowan definitely didn't understand it . . . He is a brilliant man and there is nothing there that he wasn't capable of understanding. I just don't think he was prepared at that time to put the effort into it, because of his earlier prejudices. [Crosskey, 4–5]

This kind of account is echoed by another advocate of Spencer's ideas from the laboratory where the researcher quoted above was trained.

4K

I really only started to take things seriously when we started working on ion transport and then it became increasingly obvious that there was an economy in the chemiosmotic hypothesis describing what was going on which went right across the range of what we were doing . . . so that one became convinced that this really was more likely than the other thing . . . Now the thing which convinced the world, or began to stun the world into taking notice of the Spencer hypothesis was that experiment in which he takes anaerobic mitochondria and adds a pulse of oxygen. Under those circumstances there is an ejection of protons . . . Gowan devoted himself to showing that the protons were ejected too slowly to be associated with the respiratory chain in the way in which Spencer had said. This was just to try and suppress the chemiosmotic hypothesis from another direction. But Gowan in fact never understood that hypothesis. This was very, very obvious to anyone who talked to him. He had such a dislike of it that he never bothered to think through what the consequences would be. [Burridge, 8,10]

A main point made in these passages is that Gowan got it wrong, he continued to accept erroneous ideas, because he never properly understood the chemiosmotic hypothesis. It seems to be suggested that anybody who *did* understand the theory would necessarily have accepted it. Because the correctness of the speaker's theory is taken for granted in the organisation of the account, any failure on the part of other scientists to accept that theory *must* be due to some misunderstanding. The other's failure to understand is then traced back to the action of various 'non-scientific' factors, such as undue commitment to his own model of energy coupling, a defensive attitude, prejudice, dislike and failure to put in enough effort. Thus these two accounts, like those examined above, are organised in a manner which displays how the speaker's theoretical conclusions were a simple, unmediated response to the evidence, whereas those of their opponent were influenced by extraneous or non-experimental considerations.

It is possible, in principle, that accounts 4J and 4K are closely similar because Gowan actually did not understand chemiosmosis and was defensive, prejudiced and unwilling to bother seriously with it. If one adopts this reading, there is nothing of general sociological interest about these accounts. They simply report the way things were in this specific case. However, this position cannot be held consistently with respect to *all* the accounts we have, because the accounts of the *same* actions offered by different scientists often appear to be incompatible, because the same individual can formulate significantly different accounts in different passages and because acceptance of our complete collection of accounts of error would lead to the awkward conclusion that virtually every contributor to the field, and every major contributor without exception, was scientifically incompetent and affected by non-scientific factors.

It is not possible, then, to accept at face value all the accounts of error in our material. But this is hardly surprising, for common sense tells us that scientists are likely to be sensitive about their errors and that, as a result, some of their interpretations will probably be affected by the desire to present a favourable self-image. However, neither common sense nor sociological method provide us with a way of sorting out the reliable from the unreliable accounts. As we pointed out in chapter one, not only have we no independent criteria to enable us to distinguish biased from unbiased respondents, but the testimony of each speaker generates an unending series of interpretative problems for the analyst who seeks to build up an accurate picture of what has happened in the research community.

Reconsideration of quotations 4J and 4K will help to illustrate such problems. For instance, in quotation 4J, Crosskey describes Gowan as a brilliant man who had earlier done excellent work in the field; and Crosskey notes that he himself at that time was perhaps a little dogmatic and that he has subsequently become rather more critical of certain aspects of the chemiosmotic hypothesis.[2] These points could have been used to argue that Gowan, being a researcher of great ability and much experience, may well have seen certain scientific defects in chemiosmosis as a result of which he rationally and scientifically decided that the hypothesis was inadequate or was in need of further investigation. Indeed Gowan himself, when we interviewed him, offered exactly this kind of account and buttressed it by suggesting that the judgement of *his* opponents had been swayed by non-scientific factors (4F). Moreover, several other respondents stressed that Gowan was a highly gifted scientist, scrupulous in his attention to detail and immensely industrious. There is, then, much evidence, some of which is furnished by Crosskey himself, which runs counter to the assertion that Gowan simply could not be bothered to put in the necessary effort or that he was misled by prejudice.

In addition, although Crosskey states in this passage that the chemiosmotic hypothesis was complicated, implying that this was one reason why Gowan failed to grasp it, he says elsewhere that chemiosmosis was basically quite simple. Both these apparently conflicting judgements were repeated many times in the interviews. It appears that many interviewees were, like Crosskey, able to conceive of the hypothesis as both complex *and* simple; and that they were able, at any one time, to select whichever of these characteristics fitted in with the structure of the particular account they were constructing.

Furthermore, there is the fact that Gowan can be shown to have understood the chemiosmotic hypothesis well enough to produce experimental observations which, at the time, appear to have posed very severe interpretative problems for its supporters. One of these is mentioned by Burridge in quotation 4K. Subsequently, it is said, these observations have been generally judged to be experimentally inconclusive or untenable. But speakers such as Burridge and Crosskey acknowledge elsewhere that the inadequacy of these observations was by no means obvious then and that it became established only after considerable further work by both pro- and anti-Spencerians. Thus, in other passages provided by these informants, Gowan's lack of understanding is much less obvious. In these stretches of talk, Gowan seems to have 'understood' chemiosmosis and its observational implications as well as anybody else.

We should stress that we are not trying here to disprove the accounts provided by any particular scientists, nor to show that specific accounts are intrinsically incompatible. We accept that participants, if they were given the opportunity, would be able to carry out interpretative work on their own and others' accounts to repair apparent inconsistencies. Our aim, rather, is to draw attention to the *flexibility* with which accounting is accomplished. Thus chemiosmosis was both complex and simple. It was empirically grounded, yet based only on an aura of fact. Similarly, Gowan was highly gifted scientifically yet incompetent in various respects, enormously industrious but also unwilling to make the necessary effort on a fundamental issue, putting forward criticisms of chemiosmosis which clearly showed that he did not understand the hypothesis yet which required much further experimental exploration before their inadequacy could be demonstrated.

The priority of the speaker's version of his scientific position

The two following quotations are from Wisbech, an inorganic chemist often described as being on the margins of this research network. The views proposed by Wisbech and Spencer about the nature of oxidative phosphorylation are sometimes treated as very similar, indeed as basically identical, yet on other occasions as different in various important respects. Whereas Spencer maintains that protons are transported across the inner membrane from the inside to the *outside*, thus building up a gradient which returns through specific channels in the membrane to create and set free ATP on the inside of the membrane, Wisbech argues that the protons remain *within* the membrane. Nevertheless, although the two men appear to differ considerably over most details, they can also be said to agree about the basic idea that protons and electrons become separated in the membrane and that the release of protons is essential for the synthesis of ATP. Thus, in some passages, participants treat the two men's scientific claims as essentially the same. However, Spencer's analysis is treated as having been much more influential than that of Wisbech. Spencer has written and experimented much more on the topic of oxidative phosphorylation, and it is he who has received the credit and the prizes. It is necessary to appreciate this background in order to understand Wisbech's remarks.

In the two passages below, Wisbech employs different formulations of his scientific position and, as his interpretation varies, so does his categorisation of other actors and the substance of his accounts. In 4L, Wisbech is commenting on the defects in Spencer's model, that is, on those features with respect to which Spencer differs from Wisbech. In account 4M, however, Wisbech treats Spencer's ideas and his own as identical and he contrasts 'their model' with the views of those who failed to accept the central idea which the two of them had in common.

> **4L**
>
> People were beginning to think that Spencer's hypothesis and mine were very similar. Well, the truth is they *are* similar. But the difficulty is that although both of us said that the proton and electron would escape from one another and come back and make this pyrophosphate [ATP], neither of us had a machine for doing that. Unless you invent that machine, I don't think you've solved the problem . . . And that's where the problem still is today. That machinery is not understood. In my opinion Spencer's description of that machinery is a thermodynamic impossibility, and I'm with some very good friends on that. But the biologists cannot understand why this machine is a thermodynamic impossibility. I don't believe that most of them understand this field at all . . . Spencer is an extremely naive man. He doesn't understand this thermodynamic

problem. Neither does he understand any molecular chemistry, because he's not interested in that. He's a biochemist interested in bulk levels of various things and he doesn't understand the complexity of a protein as such. So he's not a chemist. He's much more a biological man. So he *couldn't* be bothered with those properties in the least. The machine didn't bother him. The chemistry he writes down, everybody writes back immediately, not just *me*, to say, 'Well, that won't do for chemistry. The chemistry is wrong.' [Wisbech, 28,31]

4M
I said that nobody should get the prize except Spencer. (I was leaving myself out of it, because I genuinely believe I should have shared it.) The reason for that is that it was a very exciting hypothesis and his name had been associated with it. He'd worked on it when it was most unpopular, worked on it in face of a barrage of aggressive bad manners by a large number of people who didn't want it to be true, because it affected their status, I felt. And he had shown by his experiments that the ideas were basically correct. [Wisbech, 34]

These accounts both exhibit the asymmetrical structure with which we are now familiar. In both cases the speaker's version of correct belief is treated as relatively unproblematic. In 4M, it is presented as having been shown to be (basically) correct by experiments. The qualification 'basically' allows for the fact that Spencer's views are not identical with those of the speaker and cannot, therefore, have been *completely* confirmed experimentally. In 4L, those features of Spencer's theory which do not coincide with the speaker's are chemically wrong or thermodynamically impossible, or the relevant phenomena are simply not understood. The speaker acknowledges that he is offering a personal opinion. But this opinion is immediately strengthened by a reference to the 'very good friends' who endorse Wisbech's opinion; and by the subsequent observation that *everybody* objects to Spencer's chemistry and that the latter's attempts in this direction simply 'won't do for chemistry'. This portrayal of the scientific validity of the speaker's views and the support they enjoy among all competent scientists contrasts strongly with the representation of incorrect belief and the social and psychological characteristics of its perpetrators. In 4M, the latter are described as aggressive, bad-mannered and as unwilling to accept the truth 'because it affected their status'. In 4L, Spencer's errors are explained as arising from an extreme personal naivety combined with an inappropriate professional training. In addition, the perverse views of large numbers of other biologists are also attributed to their trained incompetence.

The particularly interesting feature of these two accounts, however, is the way in which, as Wisbech changes his representation of the degree of similarity between his own and Spencer's views, so the content of his accounts alters. In 4L, Wisbech begins by stating that people were beginning to think that his and Spencer's hypotheses were very similar and, indeed, that they *are* similar. But after the first two sentences, he concentrates on identifying the differences. In particular, he stresses that there is no evidence for Spencer's proposed machinery for ATP production. The rest of the passage confirms that Spencer's machinery is wrong and it explains, by reference to naivety and so on, how it is that the error has not only occurred, but has become so widespread. In contrast, in quotation 4M, Wisbech identifies himself scientifically with Spencer when he says 'I believe I should have shared the prize.' In this passage, he appears to be treating 'the hypothesis' as something which was common to them both and he goes on to describe and explain the response given to the 'basically correct' idea which they had both advocated. Consequently, there is no reference here to Spencer's naivety, to his failure to understand fundamental issues or to his narrow disciplinary perspective. Instead, Spencer's ideas, in so far as they are also Wisbech's, are simply presented as having been demonstrated by experiment. If this characterisation of correct belief is to be effective, it is necessary for the speaker to treat Spencer's scientific voice, for the moment, as (almost as) immaculate as his own. Thus, the pejorative sociopsychological characterisation is reserved, in this account, for those who opposed the essential truth embodied in the work of both Spencer and Wisbech; and the failure to recognise its validity is explained away, once again, as a result of non-scientific influences.

These two accounts bring out in a striking manner how characterisations, not only of other participants but also of scientific positions, can be varied, through appropriate selection of descriptive phrases, through selective comparison and through omission. They also reveal, once again, how the asymmetric structure remains constant even though the substance of the speaker's assertions differs dramatically from one passage to another. At the same time, they show clearly how the accounts of error furnished by a single speaker within a period of a few minutes can vary. As a result, they add further support to our previous arguments about the extent of interpretative variability and about the impossibility of using such accounts as sources of sociological evidence for the nature of social action and belief. These accounts seem to be best understood, not as providing descriptions of participants' prior actions, but as interpretative reconstructions which can portray events in many different ways, depending on the particular interpretative accomplishments in which the speaker engages in specific passages. These accounts also confirm that the crucial component in any speaker's reconstruction is the adoption of a specific version of correct belief. These conclusions will be further strengthened in the next section, where we look at accounts of error in process of construction.

Accounts of error under construction

In quotation 4N, the respondent is talking about one of the major issues under debate at the time of our interviews, that of stoichiometry. For our purposes, we can treat this issue crudely as equivalent to: 'How many protons are transported across the membrane per ATP formed?' The answer to this question has important theoretical implications, because it was frequently asserted that any figure other than two would entail major changes in the detailed mechanisms of proton movement contained in the chemiosmotic hypothesis. On the whole, the speaker talks as an enthusiastic supporter of Spencer. But on this issue he accepts that Spencer may be wrong, although personally he doubts it, and he constructs two alternative accounts of what is happening in current research into stoichiometries.

> 4N
> We are seeing many experiments done now on stoichiometry. I don't think the question is solved yet, so let's keep an open mind and let's pursue both possibilities: (a) that Spencer is right and (b) that he is wrong, in the stoichiometry matter. If Spencer turns out to be right, I think the analysis will go as follows: that we are getting a lot of people who basically understand the theory who are rushing in a little prematurely with experiments. There have been all sorts of little tiny things like how soluble is oxygen in saline and are there temperature artefacts on mixing solutions and on the electrode – technical matters which Spencer would be better on. I have never seen people do better experiments than Spencer. There are lots who are now doing as precise and beautiful experiments, but I have not seen him surpassed in this kind of detail.
> So that would be one solution. The alternative is that Spencer is being misguided by his intuition into thinking that there must be two protons because there are two electrons . . . [This possibility] does not detract, but more or less cements the chemiosmotic hypothesis. [Spencer's opponents] are showing that there *are* protons and that they *are* being translocated, and in using the criteria of Spencer, it's all complementary.
> [Roberts, 19]

As usual, there are parts of this account which seem clearly inconsistent with other respondents' views. In particular, various researchers, including several who professed to be Spencerians, maintained either that Spencer was actually a rather poor experimenter or that his experimental skills had declined in recent years; and this opinion was based, in some cases, precisely on adverse judgements of the results Spencer was getting on stoichiometry.

However, what is specially interesting about quotation 4N is the way in which the various characterisations of participants and their actions are made explicitly dependent on the rightness or wrongness of their knowledge-claims. If Spencer is right, says Roberts, then it follows that other researchers are being over-enthusiastic or in too much of a rush to produce results on a hot topic. Consequently, they are being insufficiently careful in their experiments as well as premature in their rejection of Spencer's theoretically based reinterpretation of their findings. However, he implies, we cannot know if this is an accurate characterisation of their actions until we know *who* is right. Thus, for Roberts, how to characterise the actions of Spencer's opponents does not seem to be an empirical matter, to be decided on the basis of observation and questioning of those concerned. It seems, in general terms, to be a matter of logical necessity. If Spencer is right, then his opponents *must* have been misled by *some* kind of extraneous, non-scientific influence. If Spencer is wrong, then *something* must have interfered with his normally scrupulous experimental practice.

The same technique for constructing an account is evident in the next quotation:

> 4P
>
> *Barton:* And there have been occasions when people have said, 'Oh, him' instead of, 'Oh, that.' Sometimes people have been out to prove that somebody else is wrong, rather than [*unclear*]. But I think that inevitably things were seen in that way. I've seen other fields where things have been much more bitter. But science generally does progress very well and objectively, despite the subjective element. I think there *is* a subjective element.
>
> *Interviewer:* Do you have any idea how this personal element gets eliminated?
>
> *Barton:* Only because a sufficient number of experimenters try to make the position clear. If other people are interested enough, if it's important enough, then the work will be done again or, more likely, its ramifications will be pursued. Predictions will be followed up, more experiments done, and in the fullness of time a much clearer position will become apparent. Just as happened with the chemiosmotic theory. And then, any personal rivalry will be seen for what it was, in relation to the facts, as they become more fully established.
>
> *Interviewer:* So the experimental evidence . . .
>
> *Barton:* At the end of the day solves everything [*general laughter*].
>
> *Interviewer:* Overwhelms these private antagonisms.
>
> *Barton:* That's right. [Barton, 62–3]

According to Barton, personal rivalries will only become evident for what they are when the truth has become clearly established. Such rivalries, presumably like the other forms of social distortion employed in accounts of error, cannot be definitely identified until the speaker 'knows what the scientific facts are'.

The underlying structure of accounts of error becomes visible with exceptional clarity in quotations 4N and 4P. Both accounts show clearly how speakers' formulations of error depend upon a prior formulation of correct belief. The portrayal of scientific error which we see exemplified in these accounts is a necessary implication of scientists' formulation of correct belief. It is the reverse side of the same coin. It is for this reason that Roberts and Barton are unable to decide *which* scientists have acted improperly until the scientific truth is known, *even though they already have a good idea of how those involved may have acted improperly.* In both cases, the researcher has available one or more plausible 'ready-made' accounts of error which can be applied to almost all participants and which can either be brought into play or abandoned as soon as correct belief is established. Because correct belief itself is depicted as deriving fairly unproblematically from the experimental facts, in the long run, each speaker is able to maintain that sooner or later his interpretation of others' actions will become as reliable as his knowledge of the objective realm of biochemical phenomena.

Using the contingent repertoire

It is clear from the analysis so far that whilst scientists' empiricist formulations of correct belief take a narrow range of interpretative forms, their portrayal of incorrect belief and its causes are much more varied and flexible. Not only has each speaker to be able to use his contingent repertoire to construct plausible, ready-made accounts of error to suit the circumstances of a potentially indefinite series of different individuals and situations, but each scientist has also to be able to vary his stock versions of action in accordance with the interpretative changes occurring in extended passages of talk. One feature of scientists' contingent repertoire which contributes significantly to this flexibility is the vagueness and imprecision of its terms.

The accounts of error in our collection rely heavily on notions such as prejudice, pig-headedness, strong personality, subjective bias, emotional involvement, naivety, sheer stupidity, thinking in a woolly fashion, fear of losing grants, threats to status and so on. All of these and many other similar conceptions appear in our data. Although the general drift of an account of error in which such phrases are used is generally clear enough in common-sense terms, it is very difficult to pin down their precise meaning.

For example, what exactly does a speaker mean when he describes an eminent, multi-prizewinner as naive or stupid? In addition, speakers do not hesitate to withdraw classifications of this kind during subsequent talk, if prior attributions begin to interfere with respondents' subsequent interpretative work. For example, in the following short extract the speaker simply abandons an earlier account of error as it begins to appear inconsistent with the version of events he now finds himself constructing.

> **4Q**
> Obviously, before you make an effort like that, you have to be convinced that it's going to be worthwhile in terms of both your own self-interest as well as research interest and the field itself. I think for most people it wasn't clear that that was the case. I don't think it was deliberate obtuseness or that people were really pig-headed in the sense that I might have suggested. [Richardson, 3]

There is, then, a great deal of uncertainty and conceptual vagueness about the contingent characterisations employed in accounts of error. However, this vagueness allows the speaker room to adapt and change his position as he engages in informal discourse, without having continually to repair obvious inconsistencies. Nevertheless, speakers do sometimes run into difficulties. When this happens, it is possible to observe how the very ambiguity of the initial account can be turned to the speaker's advantage.

> **4R**
> *Pugh:* There's a technology of perpetuating mythology. It's very elaborate, the system of reviewing, the way in which certain people control the meetings. If you want to write a fascinating book, I advise you to deal with the techniques by which that's done. That provides you with an absolute technique by which you can perpetuate error for an indefinite period. If you say, 'Look, I now have evidence that the Spencer model doesn't bear close examination, 1,2,3,4,5', they'll send it out to a Spencerian and he'll give you a list of things about a mile long to do and he'll wear you out. You can't win. Every experiment you do, he's got another one that you are going to have to do. He can make it impossible. But if you write it from the standpoint of a Spencerian, he'll just say, beautiful . . .
> *Interviewer:* Do you think it is true that Spencer himself had to face up to that kind of situation?
> *Pugh:* Of course. He fought another dogma and now he has become the dogma and he knows it and is not very happy about it.
> *Interviewer:* How do you think he managed to resist the dogma, so to speak?
> *Pugh:* Well it took a long time. Violent battles. And it was better. It could

explain certain things, proton gradients, which nobody had been able to explain and he introduced really revolutionary new ideas . . .

Interviewer: Do you see any signs of your own theory coming to be accepted?

Pugh: No, none. Zero. And that's because nobody is interested. They hated the problem in the first place. It's beyond them. Now they feel it's been buttoned up, they don't want to hear about it. It's the ostrich approach. But this is an abandoned generation. They will be criticised severely by the historians as unequal to the task . . . they control the means and so on and they will do it until the whole thing will be like the Emperor has no clothing. It will take 20 years to find that out, by which time they will have become Lords and Princes. [Pugh, 22, 34, 35]

Account 4R is a selection from a much longer passage. The feature that we wish to bring out is the way in which Pugh appears to change the meaning of his main 'explanatory' concepts as he proceeds. In the first paragraph, he conforms to the asymmetric pattern when he describes how false beliefs are perpetuated by the technique of peer review and so on, not only in his own area, but throughout science in the past as well as the present. The notion is then used to account for his failure to get his own theory accepted, even though, as he stressed throughout the interview, it is scientifically superior. If one takes at face value Pugh's concept of the 'absolute technique' for maintaining existing dogma indefinitely, one might think that no new ideas and, in particular, no ideas which are 'really revolutionary' like Spencer's would ever be successful.

At this point Pugh is asked how Spencer managed to overcome the dogma which preceded *him*. He deals with the interpretative task of reconciling Spencer's success with his own description of the politics of science by stressing that there was 'violent resistance' to Spencer and that it did take 'a long time'. But in order to provide some positive reason for Spencer's success, Pugh falls back on the explanatory power of Spencer's theory, even though elsewhere Pugh describes that theory as 'preposterous, unbelievable' and 'non-explanatory'. However, if Spencer's theory succeeded because it explained the experimental facts better than the chemical theory, despite the operation of the 'absolute technique', this reopens the question of why Pugh's theory, which he views as an advance on chemiosmosis, shows no signs of winning converts. Pugh deals with this in the final paragraph by returning to the control exercised by the dominant Spencerian elite. But he supplements that now-weakened notion with a series of additional factors, only a few of which appear in the quotation above.

There is clearly no point in trying to establish exactly what Pugh means by concepts such as 'technique for perpetuating mythology' or 'abandoned generation'. They have no clear referents. Nor is there much point in the listener raising what could be regarded as counter-instances, for, with slight adjustments to the account, these can always be plausibly incorporated. Like the speakers in the preceding accounts of error, Pugh employs a repertoire of interpretative resources which can be made to fit loosely but plausibly with events. These notions can be used to explain, adequately enough according to the undemanding requirements of fast-moving and unreflexive ordinary conversation, why his theory is rejected. In other words, the main characteristic of these resources is that they can easily be expanded or contracted, withdrawn or supplemented, without creating glaring inconsistencies, to meet the exigencies of each new conversational exchange. They enable speakers to carry out complex and subtle interpretative work in a way which always leaves them room for further manoeuvre and which always seems to allow the speaker's own scientific views to emerge unscathed.

Throughout this chapter we have described scientists' accounts of error as being couched in terms of that contingent repertoire which we previously observed in scientists' informal versions of laboratory practice. It may appear that we are using the notion of 'contingent repertoire' rather loosely in applying it to both these kinds of talk. It is quite possible, for example, that the lexicon for talking about laboratory practice is systematically different from that used by scientists in accounting for error. It certainly seems likely that the former topic will not feature so many references to strong personalities, to manipulation of the refereeing system, and so on; whilst the latter topic will focus much less on craft skills and intuition. However, although there may well be differences in the incidence of certain kinds of phrases in these two interpretative contexts, in both, scientists' actions and judgements are depicted as those of specific individuals acting on the basis of personal inclinations and particular social positions. Furthermore, in both topics the distinction between empiricist and contingent formulations is clearly observable, highly recurrent and recognised by participants. But the relative clarity of this interpretative boundary is highly unusual. Within the broad realm of contingent discourse, interpretative divisions are much weaker, more blurred and open to creative modification. What is distinctive about scientists' accounts of error is not so much the employment of a standard range of substantive characterisations, although such stock interpretations do seem to reappear in our material, but the pronounced tendency to organise such accounts around an asymmetrical counterposition of empiricist and contingent versions of action and belief.

Symmetrical accounts

Although the great majority of accounts of error are asymmetrical, it is not logically impossible to provide symmetrical interpretations of error and correct belief in the same account. In the next quotation, the speaker is referring to disagreements between himself and Spencer with respect to the interpretation of stoichiometry experiments.

> 4S
>
> We think we've done experiments with NEM which test the interpretation proposed by Spencer and show it's not right. He does not, presumably, believe those experiments, although he hasn't specifically said why. Instead, he offers some experiments of his own, which he would take to demonstrate that NEM is working in some different way. Again, we haven't criticised those experiments specifically to him. But we either think they're not relevant (and although he sees certain effects, they're not relevant to the problem), or he's misinterpreted things.
>
> Both camps, I think, believe that the other is emotionally involved in the answer and, therefore, there's not much point in rational argument. I don't think it's worthwhile having a rational argument with Spencer about it, because I'm fairly sure I shan't change his mind . . . I find it quite difficult to argue about this, because I cannot see how he cannot accept that our arguments and experiments are right. I suspect that he has the same problem. So I don't think it's a problem of straight science. [Beamish, 23–4]

In this account, without relinquishing his claim that he has got it right and that Spencer has got it wrong, Beamish maintains a fairly detached stance toward his opponent's views. Thus, he begins by offering a general description of two different, but experimentally-based, perspectives on the issue of stoichiometry. If he had stopped at the end of the first paragraph, we would have been left with an account which was symmetrical in the strong sense that both correct and incorrect belief were presented as scientifically legitimate. In the second paragraph, Beamish goes on to suggest that those involved do not in practice regard their opponents' work as having the same scientific status as their own. Both camps are depicted as having devised parallel, and asymmetrical, versions of what is happening. Both sides, it is said, view their opponents' interpretations as distorted by emotional commitment.

Beamish's recognition that his opponents would probably describe him in exactly the way that he describes them is an unusual personal achievement; and it is this which enables him to maintain in this passage a precarious symmetry in his interpretation of both sides of the dispute. Nevertheless, he does not deviate very far from the standard accounts of error examined above. For he also portrays participants as endorsing the strongly asymmetrical view that it is always the other side which is 'emotionally involved in the answer'. Moreover, the asymmetry is clearly linked in this passage to an empiricist formulation of correct belief. It is precisely Beamish's empiricist contentions which make it difficult for him simply to accept that all parties are emotionally committed to their own positions. In this passage, Beamish seems to treat what he calls 'straight science' as consisting of the rational appraisal of reliable evidence leading to unequivocal conclusions. Yet, on one side or the other, false beliefs persist in relation to stoichiometry. Clearly, then, this is not 'straight science' in Beamish's sense. It follows necessarily, therefore, that something unscientific is happening, that non-scientific influences are somewhere at work and that, given the validity of his own scientific views, it cannot really be the speaker who is at fault: 'I cannot see how he [Spencer] cannot accept that our arguments and experiments are right.'

The quotation from Beamish is typical of speakers who attempt to construct symmetrical accounts in that such speakers, whether they are explaining true and false belief equally in experimental terms or in psycho-social terms, tend to revert back quickly to the dominant asymmetric structure. In other words, symmetrical accounts tend to be unstable. Another example of this can be found in quotation 4H, where the speaker, having referred to the very strong individuals who want to interpret everything in terms of their theories, formulates his own empiricist practice in a humorous manner: 'Of course, those are the other guys, not us. We're interpreting it even, balanced [*laughter*].' The speaker's jokey delivery seems to imply that he is well aware that other scientists, including those he has just criticized, might also insist on presenting their views in empiricist terms and that they might also wish to explain away *his* errors as personal aberrations. To this extent, the speaker in this passage implies an equality of status amongst competing accounts and, in this sense, organises his account of correct and incorrect belief symmetrically. Yet the speaker's words, if read literally instead of being heard as a joke in accordance with his vocal inflections, remain strongly asymmetric: 'Those are the other guys, not us.' Moreover, although the speaker's tone seems to have made those words ironic at the time and to have suggested that the listeners should hear them as meaning the exact opposite of their literal sense, the speaker himself, in his very next sentence, appears to take for granted their literal meaning. He seems entirely to disregard his own humorous interjection and to proceed as if he had in no way departed from his customary empiricist voice. When the laughter ceases, he returns immediately to the theme of how other scientists seek to 'bend the data' in ways which provide spurious support for their erroneous theories. In other words, this speaker's move towards a symmetrical treatment of truth and error, like that of Beamish, is almost instantaneously abandoned.

Not only do symmetrical accounts tend to be unstable, but they are also very few in number. If we confine ourselves solely to material from our 34 interview transcripts, we find that out of a total of 65 accounts of error and correct belief, no more than five are symmetrical; whereas 60 exhibit the pattern which we have called asymmetrical accounting for error.

Accounts produced outside the interview

All the material above has been taken from transcripts of our interviews with biochemists. It is necessary to ask, therefore, whether this form of accounting occurs only in interviews. It is conceivable that there is something about the interview situation in general, our position as sociologists interviewing scientists, the particular questions we used or the inclinations of scientists within interviews which generated this kind of asymmetrical accounting. We have been unable, however, to discern any general pattern in the way in which accounts of error are produced in our transcripts or to identify any particular category of respondents as especially responsible for such accounts. Accounts of error sometimes occur early in a passage, in direct response to a question from the interviewer. But just as often they appear in the course of a more extended stretch of talk, as the speaker builds further interpretations upon his own prior discourse. They undoubtedly occur frequently in passages where 'chemiosmotic speakers' are making sense of the 'resistance' of those who are deemed not to have adopted this theory despite its experimental warrant. But they are employed with no less enthusiasm by 'non-chemiosmoticists' as they deal with the task of making understandable the apparent popularity of this 'obviously inadequate' theory. With one exception, we can find no way of consistently linking the appearance of asymmetrical accounts to variations in the form of question, the interpretative context or the type of respondent.

The only discursive feature of the interview transcripts which does seem to be clearly related to the occurrence of accounting for error is of a very general kind, namely, the predominantly retrospective character of the talk recorded in these transcripts. Because accounts of error are mainly retrospective, the retrospective orientation of interview talk probably means that the frequency of accounts of error is higher in interviews than elsewhere. But, as we pointed out earlier in this chapter, retrospection is no way confined to interviews. It is, rather, a recurrent and organised feature of ordinary discourse. Thus, if we make the reasonable assumption that retrospection is widespread within *scientists'* ordinary discourse, and if we accept that asymmetric accounts of error do not appear to arise from any unique interpretative situation being created in the interviews, then we have grounds for assuming that asymmetrical accounting will occur in the course of ordinary interchange between scientists.

The line of reasoning is strengthened by the following observations. In the first place, there is direct evidence that the same interpretative pattern is employed in social contexts other than that of the interview. It occurs, for example, in an historical article about the Pasteur effect written by one of the leading figures in the field for his fellow experts. In this article the author, quite removed from the influence of any sociological investigator, constructs an account of error with exactly the same structure as those examined above. He takes for granted that his own scientific understanding of the Pasteur effect is objectively correct and that he can see clearly what was wrong with the views of his predecessors. Because he takes the correct interpretation to be obvious, it follows that previous views were obviously wrong. He is led to suggest, therefore, that non-scientific factors were at work: 'For interesting psychological reasons this explanation of the Pasteur effect was widely accepted in spite of the fact that it was patently inadequate.'

The next quotation also indicates that asymmetrical accounting for correct belief and error does occur outside sociological interviews. It is taken from a letter written to reply to our inquiry asking whether the author would be willing to be interviewed. Our initial letter said very little about our interests and was phrased in the most general terms.

> 4 T
> I was pleased to be approached by you, especially since I have not been able to propound my own ideas in print . . . As you may infer I do not accept Spencer's various schemes and basically regard the ox phos process as arising from [technical details omitted to prevent identification]. I send a MS rejected by pro-Spencer referees but I would make the point strongly that when a body of scientists have committed themselves in print to a theory they become biased towards rejecting other ideas lest they be made to look foolish. This is a weakness of our system. I would suggest for your attention the proposal that many ordinary biochemists have been bemused and confused by Spencer's interwoven assumptions; rather than say that they do not understand them they accept them. [Sephton]

The asymmetric account contained in this letter is of interest to us here, not only because it appears outside the interview and was in no way directly solicited by the investigators, but also because there is good evidence that Sephton actually described his situation in similar terms when talking to his colleagues. In the course of another interview, a scientist who knew Sephton informally and was familiar with his work and opinions told us, without our asking, that Sephton thought he was sometimes victimised by pro-Spencerian referees but that this was not in fact so. The informant had a quite different interpretation of what was happening.

The significance of this alternative view is not that it shows Sephton to have been wrong. The most that one can conclude in this respect is that, as we have seen many times, it is quite normal for one participant's account to be treated by others as obviously unconvincing. Its significance here is rather that it shows that the response Sephton gave to us was similar to the account he had given to his colleague. It provides a clear indication, therefore, that the form of asymmetrical accounting which we have documented above does occur in the course of informal interaction among scientists.

It would be misleading to pretend that the evidence and considerations advanced in this section can lead us to more than the most tentative conclusions. For the moment, we can only conclude that the marked and dominant interpretative pattern observed in our material must occur to some as yet unknown degree in naturally occurring discourse within science; that it certainly does occur frequently in some forms of discourse involving scientists and outsiders like ourselves; and that further comparative studies of scientists' discourse are needed if we are to begin to understand more fully the social production of scientific error.

Scientists and mundane reasoning

We stressed in the last section that the restricted nature of our empirical evidence on biochemists prevents us from making strong claims about the incidence of asymmetrical accounting for error in naturally occurring situations. Furthermore, because so few studies have carried out the kind of analysis we are attempting here, there are at present no other published studies of scientists' discourse which can be used to explore the wider relevance of our observations. There is, however, one description of *non*-scientists' discourse which bears a striking resemblance to our own and which indicates that something akin to the phenomenon we have been discussing in this chapter is observable, not only in science, but also in quite different realms of social life. This is contained in Pollner's study of what he calls 'mundane reasoning about reality disjunctures'.[3]

Much of Pollner's analysis deals, not with particular kinds of social actor, but with the generic figure of the 'mundane reasoner', that is, any actor whose speech presupposes that there is an objective world which can be shared by, and identically reported by, all other competent participants. Clearly, the scientists quoted above are usually mundane reasoners in Pollner's sense. But so are non-scientists in certain types of situation; for example, people testifying in traffic courts. What is particularly interesting about Pollner's analysis is his suggestion that participants tend to deal with reality disjunctures, that is, 'disjunctive experiences and/or accounts of what is purported to be the same world'[4] in a manner which closely resembles the interpretative practice of accounting for error.

> *For a mundane reasoner, a disjuncture is compelling grounds for believing that one or another of the conditions otherwise thought to obtain in the anticipation of unanimity, did not.* For example, a mundane solution may be generated by reviewing whether or not the other had the capacity for veridical experience. Thus, 'hallucination', 'paraonia', 'bias', 'blindness', 'deafness', 'false' consciousness etc., in so far as they are understood as indicating a faulted or inadequate method of observing the world serve as candidate explanations of disjunctures. The significant feature of these solutions – the feature that renders them intelligible to other mundane reasoners as possibly correct solutions – is that they bring into question not the world's intersubjectivity but the adequacy of the methods through which the world is experienced and reported upon. *The application of such designations declares, in effect, that intersubjective validation of the world would obtain were it not for the exceptional methods of observation and perception of the persons identified as employing them.*[5]

Although Pollner's empirical material is actually rather limited, it does draw our attention to the possibility that the asymmetrical accounting for true and false belief among our biochemists is part of, or is linked to, much wider discursive regularities than at first seemed likely. Nevertheless, non-scientists' accounts, as documented by Pollner, are by no means identical with those of our respondents. For example, Pollner's subjects do not seem to employ any highly standardized devices, such as our biochemists' 'strong personalities' or 'manipulative refereeing'. This may simply mean that those suspected of traffic offences do not form a linguistic community in the way that certain groups of scientists probably do. Similarly, Pollner's subjects appear not to have any shared interpretative repertoire in terms of which they can formulate and warrant their own versions of the natural world. Pollner's mundane speakers seem to have to develop the grounds for their own claims in a relatively *ad hoc* manner out of their own idiosyncratic and defeasible experiences. For example:

Defendant: From the time I got on the freeway until when he pulled me over, I was checking my speedometer constantly . . .
Judge: You ever have your speedometer checked?

Defendant: . . . so I parked my car and I went over and I was inside their car talking for a few minutes and then the police barricaded both ends of the street off so we couldn't leave, then they charged me with aiding and abetting a drag racing contest, and there was no drag racing at all taking place.
Judge: Well, the officers appeared at the scene of extensive drag racing
. . .

Unlike these laymen, who are dealing with relatively isolated events in the unfamiliar linguistic context of the law court, our respondents are reconstructing accounts of experiences which have played a central part in their professional biographies, about which they and their colleagues have probably talked many times, and in retailing which they employ well established and appropriate interpretative repertoires. In particular, when formulating their own claims about the phenomena of the natural world, scientists have at their disposal the interpretative forms of the empiricist repertoire, which enable them to translate their idiosyncratic and defeasible experiences into the impersonal linguistic currency of 'experimental evidence'. As a result, scientific speakers seem to be peculiarly able to construct accounts in which they appear to have privileged access to the realities of the natural world: indeed, no matter what the diversity of views, each scientist manages to convey the strong impression that his voice and that of the natural world are one and the same.

It would be premature to conclude, however, that scientists' accounts of error will always be marked by a more definite contrastive pattern than those of laymen or that laymen never have access to some linguistic equivalent to the empiricist repertoire. Our evidence is too fragmentary at present to decide on these questions. For example, if each of Pollner's defendants had been recorded without the presence of the police officer who had brought the charge, as our biochemists were recorded without their opponents being able to overhear them, a significantly different series of explanatory accounts might well have been obtained, perhaps closer in their structure to our scientists' interpretations. It seems more reasonable, therefore, to conclude this chapter by stating that, so far as we can judge from our data, scientists' accounts of error appear to have a well-defined interpretative structure, that this structure depends on the existence of the two repertoires previously identified, and that there is some evidence of somewhat similar structures occurring in one other area of discourse.

5

••

The truth will out

Throughout the preceding chapters we have emphasised the variability of scientists' discourse. So far, however, we have not considered how, despite the marked degree of interpretative variation, scientists manage to maintain an adequate appearance of consistency. How is it that they manage to proceed without constantly generating apparent contradictions? Let us take this question as our point of departure in the present chapter. In order to make it more concrete and, therefore, more amenable to a clear empirically-based answer, let us reformulate the question in terms of the two interpretative repertoires identified above. Thus the question becomes: 'If scientists regularly draw upon and move between two quite different repertoires, how is it that potential contradictions between these repertoires do not require constant attention?'

As we saw in chapter three, when the repertoires *are* brought together by scientists, they tend to be treated as distinct and as incompatible. For example, we observed in that chapter how scientists, when discussing their own papers informally, sometimes noticed that the accounts they gave of laboratory practice in research reports were significantly different from the accounts they gave in interviews and we examined the attempts of two respondents to explain what they took to be inconsistencies between the two kinds of account. Thus we showed in chapter three that the juxtaposition of accounts involving the two repertoires does tend to generate interpretative work which focuses on potential incompatibilities. But we also showed that the contingent repertoire is largely excluded from the formal research literature and that, by implication, this contextual segregation of the empiricist repertoire serves to reduce the frequency of potential interpretative contradictions.

In informal talk, however, both the repertoires are regularly employed and the likelihood of interpretative inconsistency is accordingly increased. But even within informal talk there is a tendency for speakers to keep their use of the two repertoires separate. Informal talk proceeds with great rapidity and a pause of as little as a second is typically noticed as a significant hesitation.[1] Hence, the time lapse between different versions of action does not have to be particularly long for them to be heard as separate within ordinary conversation. Nevertheless, scientists' two interpretative repertoires are sometimes intimately combined during informal discourse. We have seen an example of this in the case of accounting for error. However, interpretative inconsistency is avoided there by the very structure of the account, each repertoire being applied systematically to distinct categories of action and belief.

These observations seem to imply that interpretative inconsistency may not be such a pervasive problem as might initially have been imagined. The structural features of informal discourse contribute to the meaning of its components in particular passages and tend to reduce the likelihood that closely juxtaposed elements will be heard as irreconcilable. Moreover, as we stressed in chapter four, the very flexibility of much informal discourse helps speakers to cope with the potential contradictions generated in their own speech. However, the preservation of consistency is not an automatic process; nor is it an inevitable outcome of participants' interpretative work. Indeed, our discussion so far enables us to identify the precise kind of situation in which interpretative consistency is most at risk. It will be most at risk in situations where, as in accounting for error, the two repertoires are closely combined in one continuous passage, but where, *un*like accounting for error, both repertoires are applied to the same events or to the same class of events.

In the light of our prior analysis, we would undoubtedly expect scientists to experience some kind of interpretative difficulty in circumstances of this kind. Yet, given the flexibility of everyday interpretative resources, we would not expect them to find these difficulties insuperable. What we would expect, and what in fact we find, is that when the two repertoires occur together in the course of ordinary talk, other than that of accounting for error, their potential incompatibility is often signalled by, as well as being resolved by, a specific interpretative pattern or device. We call this the 'truth will out device' (TWOD). In the rest of this chapter, we will concentrate on the TWOD. Our aim will be to show that our form of analysis and our earlier analytical conclusions can be extended in a way which improves our understanding of some of the fine detail of discourse in science. We suggest that the kind of data to be examined below would have no significance for other approaches at present available in the sociology of science. Although these approaches rely on scientists' discourse as the basis for their own analytical claims, they provide little in the way of detailed insight into the organised character of the discourse.

The TWOD is by no means such a recurrent device as asymmetrical accounting for error. Whereas the latter occurs more than 60 times in 34 transcripts, we have at present only ten cases of the TWOD. Nevertheless, the pattern is systematic and strongly linked to the coming together of the two interpretative repertoires. In the following sections we will examine in detail three instances of the 'truth will out device', and a fourth closely similar pattern. We expect that the TWOD will be one of a family of 'reconciliation devices' arising from scientists' movement between interpretative perspectives. We hope that its identification, as well as strengthening and extending the conclusions presented in previous chapters, will stimulate other researchers to look at their data with the possibility of such devices in mind. Because we are going to examine our material in greater detail than in earlier chapters, we have numbered the sentences in each quotation for ease of reference.

An example of the 'truth will out device'

The following quotation from an interview transcript illustrates clearly how the TWOD, which is formulated succinctly in the final sentence, is used to resolve interpretative difficulties arising from the speaker's use of a contingent as well as an empiricist repertoire. Before this passage the scientist concerned had been talking about his own and other people's adoption of the chemiosmotic hypothesis. He talked exclusively of the theory's capacity to make consistent sense of experimental evidence. He was then asked whether there were any general criteria which could be used in choosing between theories. He begins his answer by citing a criterion which, he says, was favoured by the originator of the chemiosmotic theory.

> 5 A
> 1 Spencer was always very fond of Occam's Razor and that's as good a starting-point as any, I think. 2 If you can get away with a few hypotheses, then it's useless having more. 3 It's not a guarantee of being right, of course. 4 Well, I think in the long run it *is* a guarantee of being right, in the sense that if you add everything together it's really the only basis for believing a hypothesis. 5 In fact, when you're struggling with a compendium of facts and personalities, it's a useful thing to bear in mind certainly. 6 But I think it would be dishonest to say that that's how it actually works. 7 One grows very much aware of science as a *social* system, especially in this area where things have never been very clear-cut and where there have always been areas in which the evidence itself was conflicting. 8 To quite a large extent, one's based the path one has followed as much on intuition, that is, a feeling that it's right, which obviously is only useful if you've got a pretty large body of working evidence mulling around in your head at the time. 9 So intuition based on experience and, secondly, on one's feeling for the honesty and capacity of one's collleagues. 10 I think that's a very important thing, which is very often unsaid in science, but I'm sure plays a very important determining

role in the progress of ideas. 11 I know it's so in scientific meetings. 12 People will pay attention to some people and not to others. 13 And sometimes it's a very false sort of thing, because it also has mixed up in it the whole thing of charisma and how nice a person is rather than how competent they are. 14 So it's a somewhat unreliable guide, but I'm sure it plays an important part in determining the course of events. [*Pause*] 15 I think *ultimately* that science is so structured that none of those things are important and that what is important is scientific facts themselves, what comes out at the end. [Richardson, 12]

The speaker's use of an empiricist repertoire for talking about theory-choice largely precedes this passage. However, in this quotation, he begins in traditional fashion with Occam's notion of the right theory being the one which rationalises the evidence most economically. But then he wavers. Contradictory statements are proposed as to whether or not this criterion is a *guarantee* of being scientifically correct. He first of all says that *of course* it *cannot* be a guarantee. He then reverses this opinion and claims that *in the long run* it *is* a guarantee. This reference to 'the long run' prefigures his eventual use of temporality in the TWOD. However, at this juncture he does not develop the idea further. He returns to a conception of a scientist trying to choose between theories at one point in time, and once again he revises his prior statements about the certainty provided by Occam's Razor. In this third attempt to say how it operates, he ends up with a very weak formulation indeed. The principle of economy becomes merely something to 'bear in mind' when one is 'struggling with a compendium of facts and personalities' (5A5).

He then says that 'it would be dishonest to say that that's how it actually works' (5A6). It is unclear which of his previous claims is being modified here. It seems unlikely, however, to be his third interpretation of Occam's Razor, which could hardly be weakened any further (5A5). Thus this sentence (5A6) seems to be an additional statement designed to correct his strong assertion that Occam's principle is a guarantee of truth (5A1–2). What we find, then, in the first six sentences of this passage is an oscillation between strong and weak assertions about the certainty furnished by a major criterion for assessing scientific theories. This opening sequence ends with the strong claim having apparently been abandoned and with the suggestion that the establishment of scientific knowledge involves the assessment of researchers' personalities as well as the explanation of biochemical facts. In terms of our analytical categories, this sequence shows the speaker moving hesitantly away from an empiricist conception of scientific action and belief, according to which relative certainty and factual substantiation are the dominant features of scientific knowledge, towards a more uncertain, contingent perspective, in which social and personal factors are allowed to have a significant influence on scientific thought.

Sentence 7 begins to develop a strongly contingent account of how science in general works, but this is immediately qualified by reference to the special experimental conditions found in bioenergetics (5A7). Nevertheless, he does appear to generalise his point to include to some degree the whole of science. The following sentence is also organised in terms of a strongly contingent assertion being modified by an empiricist qualification. The speaker begins by stressing that he has had to base his scientific judgements on such uncertain and idiosyncratic elements as feeling and intuition, but he then reintroduces the notion of experimental evidence and continues by emphasising that even intuition can be based on experience (5A8). Thus both sentences 7 and 8 begin with what looks like a strongly contingent assertion, which is immediately qualified by making the operation of the contingent factor to some unspecified degree dependent on experimental evidence.

In sentences 9 to 14, however, the speaker identifies a strongly contingent element, namely, judgements about honesty, capacity and charisma, which he says play an important role in determining the progress of ideas. This is the most sustained, purely contingent sequence in the whole passage. Furthermore, it does not explicitly refer to the possibly special case of bioenergetics. The speaker seems to be offering a contingent account of scientific action and belief in general which, he says twice, he is sure is correct. This sequence, unlike previous references to contingent factors, is not internally qualified.

It is immediately after this sequence that the speaker introduces the 'truth will out device'. What this device does is to reintroduce the time element, which had been briefly mentioned in sentence 4. All that has been said before is now to be seen in terms of a temporal development. Gradually, it is implied, the realities of the physical world will be recognised; and idiosyncratic, social, distorting influences will consequently be seen as such (5A15). The speaker's statements about the importance of contingent influences on scientific thought are not withdrawn. Social factors, personal judgements, intuition, charisma and so on are all allowed to play a part in science. But only in the short run. In the long run, it is scientific *facts* which are important. Thus the TWOD resolves any apparent contradiction between the speaker's constant reference previously to experimental evidence as the sole basis for theory-choice and his subsequent account of contingent factors, by separating the empiricist from the contingent elements over time. Both his empiricist and his contingent statements can now be taken as correct because, it is now implied, they referred all along to different phases of scientific development. Although empiricist factors are depicted as operating throughout the temporal sequence during which scientific knowledge is produced, they are treated as becoming increasingly effective over time. In contrast, contingent factors are portrayed as being influential initially, but as dropping away over time. The TWOD thus enables the speaker to re-establish an interpretative disjunction between the two repertoires after they have been intimately combined in one conversational passage.

In the full transcipt, the use of the TWOD is followed by a partial change of topic. The speaker does not enlarge further on exactly how contingent factors are eliminated or how facts come finally to assert themselves. There is no attempt to specify the structure of science mentioned in sentence 15, which ensures that the truth *will* out. Instead, the speaker goes on to consider what *might* have happened if Spencer's theory had actually been introduced by a scientist already eminent in the field, rather than by an 'outsider'. He suggests that it would have been accepted much more rapidly. Thus once the TWOD has been asserted in this sequence, but in no way substantiated, it seems to be taken for granted in formulating the story which follows. It is assumed, for instance, that the same theory would have succeeded in the end. There is no consideration of the possibility that this other scientist might have developed 'the theory' differently and might have used his social position to gain its acceptance. Social and personal factors are no longer treated as playing 'a very important determining role in the progress of ideas'. The introduction of the TWOD enables the speaker to present these contingent elements as if they had merely a minor accelerating or retarding effect on the acceptance of the one truth. Clearly, not only does the TWOD resolve potential contradictions between the empiricist and contingent repertoires, but at least in this instance it works to re-establish the primacy of the empiricist perspective and to confirm the inevitable predominance of the speaker's own theoretical (in this case, chemiosmotic) views.

The TWOD as an interpretative accomplishment

So far we have referred to the 'truth will out *device*', even though we have given only one example. In order to deserve the term 'device', this pattern of interpretation needs to occur regularly, to be closely similar in its internal construction on different occasions of use and to be a response to an identifiable interpretative context. Let us look at another passage from a different transcript in order to begin to show that these requirements are satisfied. This quotation is taken from the very end of the interview, when the interviewers asked a final, open-ended question.

5 B

Interviewer: 1 I think that's all. 2 Perhaps we could ask one further question. 3 Are there any things which you think are important about this field, but which we haven't touched on?

Barton: 4 Yes, you haven't touched on personalities *very* much. 5 Spencer and so on. 6 I'm not sure I want to talk about them. 7 But I think they *have* contributed.

Interviewer: 8 Would you say something general without naming names?

Barton: 9 The thing of which I'm well aware is that the attitude that Mulhern took to anything Burridge published, which was of severe, critical, bitter opposition. 10 He didn't like him. 11 His bitterness has disadvantaged he, Mulhern, enormously. 12 Because it meant that other people distrusted his judgement. 13 And there have been occasions when people have said 'Oh, him' instead of 'oh, that.' 14 Sometimes people have been out to prove that somebody else is wrong, rather than [*unclear*]. 15 But I think that inevitably things were seen in that way. 16 I've seen other fields where things have been much more bitter. 17 But science generally does progress very well and objectively, despite the subjective element. 18 I think there *is* a subjective element.

Interviewer: 19 Do you have any idea how this personal element gets eliminated?

Barton: 20 Only because a sufficient number of experimenters try to make the position clear. 21 If other people are interested enough, if it's important enough, then the work will be done again or, more likely, its ramifications will be pursued. 22 Predictions will be followed up, more experiments done, and in the fullness of time a much clearer position will become apparent. 23 Just as happened with the chemiosmotic theory. 24 And then, any personal rivalry will be seen for what it was, in relation to the facts, as they become more fully established.

Interviewer: 25 So the experimental evidence . . .

Barton: 26 At the end of the day solves everything [*general laughter*].

Interviewer: 27 Overwhelms these private antagonisms.

Barton: 28 That's right. [62–3]

Once again, we have not quoted the extended sequences in which this scientist talked exclusively within an empiricist repertoire. As with almost all our respondents, this happened repeatedly throughout the interview. Indeed, this speaker takes as his point of departure, in sentences 4 to 6, the relative absence of prior references to the contingent factor of 'personalities'. He identifies 'personalities' as something which has contributed to the field in some, as yet unspecified way, which he *could* talk about, but which he would prefer not to talk about. He seems to imply that unless the interviewers do come to appreciate the part played by personalities, they will not fully understand the nature of the field (5B6). The speaker is

drawing attention here to phenomena which are important for participants, which exist for participants only in so far as they are constituted
through use of the contingent repertoire and which can only be described
and understood in terms of that repertoire.

Like the speaker in quotation 5A, this scientist hesitates at the outset. He
has to be encouraged by the interviewer's sentence 8 before he begins to
deal concretely with the impact of personalities. Then, despite having been
invited to speak in general terms, he starts to talk about the supposed
strong personal antagonisms of particular scientists. These specific
statements lead to a series of increasingly strong and general assertions.
Personal bitterness is said to have influenced scientists' knowledge-claims
(5B9–10). This is taken to be inevitable; that is, presumably, it is an
endemic feature of science (5B15). Moreover, personality clashes in other
fields are said to have been much worse. Bioenergetics, it seems, is by no
means a special case (5B16). Thus, in the course of a few sentences
employing the contingent repertoire, Barton seems to have come close to
undermining the basic assumptions of much of his previous discourse.
However, his strongest contingent claim in sentence 16 is immediately
followed by a reaffirmation of the empiricist position and by a reference to
temporality (5B17). The sequence then ends with a return to the topic of
contingent factors. We are told that his references to subjective factors are
justified, but that their existence is in no way inconsistent with the
objective progress of science (5B18).

At this point, the TWOD has been strongly implied, but not clearly
expressed. The interviewer then replies by adopting the respondent's
perspective and by asking for further clarification. He takes over the
notion of a progressive move towards the truth from sentence 17, develops
this in terms of the elimination over time of personal influences, but poses
a question for Barton by treating the details of his empiricist assertions as
insufficiently explicated (5B19). Unlike the speaker in 5A, this scientist
is now obliged to do more than furnish a simple version of the TWOD. For
this has effectively been accomplished already through the interaction of
interviewer and respondent in sentences 17 to 19. Consequently, he responds in sentences 20 to 22 by describing succinctly how reliable experimental evidence is actually established. He then refers back to the case of
the chemiosmotic theory and makes it clear that the effects of the personal
antagonisms he had just described did not prevent the eventual adoption
of the right theory in this case. Finally, in sentence 24, he produces a clear
general formulation of the TWOD and, with some prompting from the
interviewer, he extends this version to make it so all-embracing that it
provokes outright laughter from all concerned (5B26).

In quotations 5A and 5B, we can see clearly that both scientists are faced with a similar interpretative problem. They have both made extensive use of an empiricist repertoire in describing what goes on in science. They both spontaneously, albeit rather hesitantly, begin talking about the importance of contingent factors in such a way that, despite their continual reservations, scientists' actions and judgements come to appear rather uncertain and arbitrary. They both then deal with possible contradictions between their two kinds of statements by offering very similar formulations, according to which empiricist and contingent elements become separated out over time. 'I think *ultimately* that science is so structured that none of these [personal] things are important and what is important is scientific facts themselves, what comes out at the end.' 'In the fullness of time a much clearer position will become apparent . . . and then, any personal rivalry will be seen for what it was, in relation to the facts, as they become more fully established.'

The TWOD from quotation 5B is particularly interesting, because it suggests that personal rivalry, and presumably contingent factors generally, are actually identified as such by reference to the supposed scientific facts. Thus the speaker takes for granted that his recognition over time of certain scientific claims as factual and of others as non-factual enables him to discern how personal factors are inferred from what the speaker takes to be the degree of correctness of the claim. We have, of course, observed this phenomenon before in the construction of accounts of error. If this is generally the case, if scientists' accounts are normally constructed in this way, the separation of contingent and empiricist factors, which scientists present as an observable feature in the social world of science, is, rather, a necessary consequence of their methods for constructing accounts of that social world. If each speaker identifies personal and social influences as always or almost always leading to false claims and to incorrect observations, and if he treats what he now takes to be correct claims as free of such influences, it will follow *necessarily* that currently known facts will appear to have been independent all along of personal and social influences. For contingent factors are by definition, or more correctly by interpretative procedure, those which are associated with false belief and error, and which therefore can have had no impact on what is at present known to be true. In contrast, those actions and judgements which are associated with 'the facts' will now appear, *because of* their link with 'the facts', to have been non-contingent, that is, to have been an unbiased recording of the realities of biochemical phenomena.

It is reasonable to suggest, then, that the temporal separation of empiricist from contingent factors in the TWOD is an interpretative accomplishment on the part of the speakers and not an aspect of scientists' experimental and theoretical actions themselves. This conclusion is more than an analytical suggestion derived from one turn of phrase in a single TWOD. For it also follows directly from our analysis of accounting for error in the previous chapter and from what we referred to there as the fundamental principle of social accounting in science. In other words, we are suggesting that the TWOD is another device which scientists use to construct and sustain interpretations of their social world that are consistent with empiricist formulations of their own scientific views.

It is clear that the TWODs examined so far have been constructed from an empiricist perspective. That the facts, in due course, speak for themselves and are in some way independent of personal and social factors seems to be taken for granted in each TWOD. It seems to be assumed that some facts at least can be distinguished from non-facts in an unequivocal way at certain points in time. This is implied in the claim that over time a clearer and clearer picture emerges, making possible the gradual identification and elimination of contingent influences. Moreover, it also seems to be assumed that contingent influences tend to distort scientists' conclusions and to lead to scientific error. Thus TWODs enable speakers to re-establish an empiricist basis for their talk because the organisation of TWODs takes for granted an empiricist position.

It is notable that the TWODs in our material tend to occur at the end of a speaker's conversational turn or to be immediately followed by a change of topic (seven instances at the ends of turn, one followed by a change of topic, and two others). Speakers do not normally proceed to elaborate them further. TWODs are a powerful reconciliation device because they re-state the self-evident in an appropriate fashion. TWODs enable the speaker to reaffirm the scientific legitimacy of his position where this has been put in question by his own speech. They do this by making the association between contingency and false belief a matter of interpretative fiat, by separating contingent and empiricist elements over time, and by reasserting the eventual dominance of the speaker's own empiricist formulations.

Temporality as an interpretative device

Let us examine another passage of interview transcript in order to see how far our conclusions so far are confirmed and in order to explore further the structure of the TWOD.

> 5 C
> *Hawkins:* 1 I think [science] could do with a great deal fewer theories, much less determinedly expounded theories. 2 If you look in the literature right now, you can easily see where much space is devoted to each person trying to defend their particular theory or their particular

mechanism, rather than to the experiments to show that they've got [*unclear*]. 3 I think you'd find that immediately in the journals. 4 There's an enormous number of papers published that say nothing more than 'My theory is great and here's the reason.' 5 And what good do they do? Very, very little. 6 If the theory's right, don't worry. It'll have its day. 7 If it's wrong, then you're kidding yourself and everybody else in the process and making it difficult to do good science.

Interviewer: 8 Do you think that, in this field, your views are somewhat unusual? 9 Because there *has* been, appears to have been, a lot of controversy and discussion of theories?

Hawkins: 10 Oh, well I would say that they spend a very inordinate amount of their time on theories and much too little doing experiments. 11 One fairly good evidence of that, since you're doing this you should make a quick survey of the number of meetings held per year, by the people in this field and the list of people who attend them. 12 You would find that the same group of people talk to each other fifteen times a year at various meetings. 13 I *know* they're not saying anything intelligent fifteen times a year that they couldn't tell each other once a year much more effectively. 14 So what are they doing for the rest of the time? 15 Well, the answer is they're playing politics and they're talking theories. 16 One-upmanship and so forth. 17 They're not, it doesn't [*unclear*].

Interviewer: 18 This raises an interesting question. 19 If a field is like that, can you assume that the *right* theory, observationally based theory, is going to win out in the long run?

Hawkins: 20 In the long run, yes. It may not be very soon. 21 You see, facts in the long [run], there is one thing about science, there's no way to avoid the facts for ever. 22 Eventually, whatever theories exist, they will evolve to whatever is something that will fit. 23 So you see a situation in which they may swing completely. 24 You may have one year and you can say 90 per cent of the people are in chemical coupling. 25 That was not too long back. 26 Now you can say they are 90 per cent chemiosmotic. 27 Well that doesn't mean that one or the other was right, or anything of the sort. 28 But time will tell. 29 Because eventually data will be generated. 30 I think it merely slows down the process. 31 But I have great faith, in fact, that *eventually* we will know what's going on and that's all the question really is. [12–13]

This quotation from Hawkins' transcript follows an extended and forceful exposition of the empiricist position as a scientific ideal. The speaker condemns *all* the available theories in the area for being too far removed from the facts. In addition, theories are criticised for leading to improper experimental design. He suggests that the existing theories only *appear* to be supported by the evidence because experimenters have been unable or unwilling to separate their actual data from the speculative theoretical notions to which they are committed. He continues this theme in the opening sentences of the quoted passage, applying it here to science in general. He maintains that a great deal of the scientific literature is taken up by self-interested and speculative claims which are not properly grounded in experimental evidence (5C2–4). He suggests in sentences 5 and 7 that this work is not contributing to genuine scientific knowledge. Theories are presented as being proposed and accepted for scientifically improper reasons.

The general message of this passage, then, is that much of what passes for scientific knowledge is merely contingent and incorrect. However, this negative evaluation is to some extent countered by sentence 6: 'If the theory's right, don't worry. It'll have its day.' This is a very condensed formulation of the TWOD. It introduces the temporal perspective and it guarantees that the unhappy state of affairs he has been describing cannot continue indefinitely. By means of this remark (5C6), the speaker justifies his own insistence on refusing to engage in what he treats as premature theoretical work. Although many others in his field concentrate on unprofitable speculation, it is his approach of focusing on the data which is seen as eventually leading to the truth. Thus, in this initial sequence, the TWOD operates to resolve any apparent contradiction between the speaker's empiricist account of his own actions and his contingent account of the research network to which he belongs.

At this point, the interviewer picks up the implied contrast between the speaker's account of his own actions and those of his colleagues. In sentences eight and nine he re-states this contrast explicitly, suggesting that the respondent's approach to science seems exceptional in this area. Hawkins responds, in sentence 10, by repeating his previous criticism of these other scientists and then by going on to furnish further 'evidence' of their unscientific behaviour. He asserts that all of these people meet together fifteen times a year, that they cannot need to meet that frequently in connection with genuine scientific issues and that it necessarily follows that they must be engaging in activities which are improper by empiricist standards (5C13). He concludes his turn by identifying these supposed activities in contingent terms as 'playing politics, talking theories, one-upmanship and so forth'; and by contrasting them with what scientists should be doing. Although sentence 17 is not completely audible, it is clear from the rest of the passage that they should be engaged in doing proper experiments.

The interviewer's next response is to formulate explicitly the potential contradiction between the empiricist and the contingent elements in Hawkins' interpretations of scientific action. In doing this, he refers back to Hawkins' prior assertion that the right theory will undoubtedly have its day (5C6) and proposes that this seems to be inconsistent with his depiction of the actions and judgements of many people in the field. Surely, the interviewer suggests, in these circumstances one cannot be certain that the truth *will* out. The respondent counters, however, not by specifying in any further detail *how* he can be so sure that the right theory will win, but simply by restating the TWOD in stronger form (5C20–2). In particular, he emphasises the element of temporality. The central point of this revised version of the TWOD seems to be that the time period involved may be very long indeed. Virtually every sentence places great stress on the need to give the correct theory enough time to emerge: 'in the long run', 'it may not be very soon', 'in the long run', 'no way to avoid the facts for ever', 'eventually theories will evolve that fit', 'time will tell', 'eventually data will be generated' and *'eventually* we will know what's going on'.

This emphasis on temporality is combined (5C23–7) with a strong reassertion of the contingent character of much scientific belief. Even though 90 per cent of people in the field may once have adopted the chemical theory and even though 90 per cent of people may now support chemiosmosis, this does not mean that either theory is right. Thus Hawkins stresses that even widely accepted theories may be wrong and, by implication, that their general adoption is due to the kind of personal and social factors that he has previously identified. This is one major reason why he has to extend the time period available for the correct theory to succeed. His rejection of all existing theories, in the face of an apparently firm scientific consensus, requires him, if he is to apply the TWOD to his own situation and thereby to justify his own research practice, to focus on the long period of time sometimes involved. In the *short* run, as he says immediately after this passage, there is no likelihood that a theory which he would accept will come to predominate.

This example of the TWOD provides further illustration of all the main points previously made. It grows out of the speaker's use of the two interpretative repertoires. It is employed to resolve potential contradictions. In this case, the interviewer's explicit formulation of a possible contradiction elicits a very strong example of the TWOD, which is closely similar in its organisation to those examined previously. This TWOD also resembles those above in the way in which an empiricist perspective is taken for granted and in the way in which the speaker identifies contingent factors by reference to what he takes to be the experimental facts. The respondent seems to treat the theoretical interpretations of others as self-interested speculation and their actions as contingent in various ways, simply because these actions and associated beliefs do not coincide in their scientific import with his own. Thus the speaker's assertion that other scientists are merely 'playing politics and so on' is directly derived from his opinion of their scientific views (5C13–16). It is his 'knowing' that their meetings are scientifically unprofitable which, from his empiricist perspective, means that they *must* be engaging in scientifically improper, and ultimately irrelevant, activities.

The most striking and novel feature, however, of this particular TWOD is the respondent's indefinite extension of the time element. It becomes clear from this example that speakers can deal plausibly with any situation where 'the facts' have not yet emerged, simply by deferring the time when contingent factors are to be eliminated. The eventual separation of contingent from empiricist elements is taken as an article of faith, as Hawkins himself says in the concluding sentence. The assumption that it will happen can be used to fashion scientists' accounts, without speakers being required to present the slightest evidence that such a separation has yet begun. Thus the TWOD is a highly flexible interpretative mechanism. Those speakers whose own scientific views have come to be accepted will be able to 'show' that, as the truth has emerged, so the influence of contingent factors has now been erased. This conclusion will follow necessarily from scientists' procedure of treating 'the facts' as by definition independent of such influences. However, those whose views have been less successful will not be forced to adopt a contingent view. For they will always be able to defer the elimination of contingency into the future. In short, the TWOD can be constructed in a manner which allows any scientist who is claiming that his views are correct to reconcile the two interpretative repertoires in favour of the empiricist perspective and, thereby, to anchor his own scientific views in that perspective.

The TWOD and the contingent repertoire

We have so far examined three examples of the TWOD and we have observed a very similar interpretative pattern. This pattern is clearly repeated in all but one example from our collection. In this one, a pattern similar, but not identical, to the TWOD is used in the context of a strongly *contingent* position. Let us look at this exceptional case and consider its implications.

5 D

Cookson: 1 I felt at that time that Spencer was wrong. 2 I was trying to say that if he is in fact wrong, you can't disprove him. 3 That's essentially what I was saying. 4 I wasn't really, it was a very personal kind of statement. 5 It was taking the position that I felt that there was no way that he could be right, then it was impossible to disprove him, more or less. 6 I don't think it has – well it turned out to be that I was wrong and [Spencer] was right. 7 I basically just feel that, I am so, I think that the chances of being right in interpretation are so tenuous in this business, that I think you have to really keep an open mind as much as possible and try not to become biased towards a particular model. 8 And that's essentially my feeling about this business really.

Interviewer: 9 But the reason for that is what, that none of the theories really generate very precise observational predictions, so that you can never . . .?

Cookson: 10 Yes. The thing is that you can't, it's very hard to get your hands on these things that you are working on. 11 Membranes are extremely complicated and it's hard to know that you've ever got the variables all pinned down, so that when in fact you make an observation that that observation is really what you *think* it is. 12 I am trying to think of a good example. 13 All the alpha beta stuff [terms altered to protect the speaker's anonymity] is perfect. 14 By any criteria, up till a year ago, I don't think anybody reading that stuff would have had any argument that those things bound to the membrane but weren't transported.

Interviewer: 15 Because you'd gone through all the controls?

Cookson: 16 As much as we could. 17 Any thing that made, well I frequently used to say to the guys, 'Look you know', and especially when Stephenson first challenged this paper, I'd say to the fellows in the lab, 'Look, he can't be right, we've done everything we can think of.' 18 He did something we didn't think of, OK 19 That's I think very pertinent to what we are talking about. 20 You just don't, I mean what happens with science is that you make, even when you think somebody's right, OK you think you are right or you think somebody else is right, it's going to be 10 or 15 years. 21 I think the burden of proof becomes a time phenomenon. 22 It may be very clear right now that Spencer's right, but how do you know that five years from now somebody else isn't going to come up and say 'It looked good, but it's not really protons, it's a quark pump.' 23 That's one of my favourite gags, a quark pump. 24 How do you know it's not a quark pump? 25 It's got to stand up for a period of time. 26 It's got to stand up, especially in something like this, you have to do, you can't grab, you know you can't get it, you can't catch the variables that well. 27 I mean all you can do is make a lot of observations and if they are all consistent, eventually if nobody does something that absolutely shows that it can't be right, then it's accepted as proof. [42–3]

28 It's a very complicated business, there are no set rules. 29 I mean rules change from day to day, that's the thing. 30 What bothered me with this [alpha beta] episode was the final and complete realisation that there is no such thing as absolute truth. 31 I mean, last summer, it really hit me like a ton of bricks that truth is simply what most people are willing to believe today. 32 And that's truth. 33 Tomorrow the population changes, people are not willing to believe the same stuff that they were willing to believe the day before yesterday, then truth changes. 34 And that is what this business is obviously all about, you know, clearly. [Cookson, 49]

The first thing to consider here is whether this quotation does actually contain a TWOD. The TWODs discussed above:

(a) stress the element of time;
(b) treat 'proof' as something that necessarily emerges over time;
(c) are used to reconcile the two interpretative repertoires.

The last of these three characteristics has been put forward as an empirical claim and the possibility of there being TWODs which are not so used must be left open. Thus TWODs must be identified solely by reference to features (a) and (b) above. The most likely candidate for being a TWOD in quotation 5D is to be found in sentences 5D20–7. However, although these sentences certainly emphasise the temporal element, they do not assert that 'the truth will out'. They do not propose that correct belief will *necessarily* emerge over time. The establishment of correct belief is instead treated in sentence 27 as conditional: '*If* they are all consistent, *if* nobody does something, *then* it is accepted as proof.' Consequently, this passage does not have the same structure as the previous TWODs, even though there is some resemblance. Nevertheless, despite the fact that this is not a fully-fledged TWOD, there is sufficient similarity for us to benefit from considering its interpretative organisation.

In this quotation, we have brought together two passages separated in the interview by several minutes of conversation. During the intervening period those involved continued to deal with the topic of assessing knowledge-claims and establishing proof. Then in the passage beginning with sentence 28, Cookson refers back to the alpha beta episode which had been a central topic in the previous passage. Thus his general deliberations on proof in the second passage can be seen as a gloss on the phrase 'accepted as proof' which ends the first passage.

Cookson's statements in the second passage are strongly contingent. Scientific truth is taken to be simply what most people happen to believe at any particular time (5D33). Scientific belief is presented as being strongly influenced by the people involved (5D31). The idea that there might be one final, correct set of beliefs is firmly rejected (5D28–34). In the first passage, in order to show that 'proof is a time phenomenon', Cookson suggested that Spencer's idea of a proton pump may one day be replaced by the idea of a quark pump. (The fact that this example is humorous does not affect the structure of the interpretation offered by Cookson.) But according to the second passage, it cannot be concluded that the quark pump has been proved to be better than the proton pump on the basis of any invariant rules of proof. For 'there are no set rules' and 'the rules change from day to day' (5D28–9). Thus from the strongly contingent perspective of this second passage, all he can mean by this example is that one truth has been replaced by another truth. The notion of 'proof as a time phenomenon' can only mean that at one point in time a certain theory can be proved to be correct, and at another point in time another theory can be proved to be correct. It is impossible to distinguish separate phases, in one of which a whole class of influences on scientific belief has been eliminated. The most that a speaker like Cookson in the second passage can consistently propose is a temporal sequence in which one contingent set of beliefs is replaced by another contingent set of beliefs. He cannot identify a stage of intellectual development which provides a final interpretative resting-place and in relation to which his own beliefs can be conclusively justified. Thus this quotation shows that the adoption of a strongly contingent conception of truth makes the TWOD unworkable.

We have identified the second passage as formulated within a contingent perspective. Turning to the earlier passage (5D20–7), there are several signs that an empiricist position is being used. The speaker constructs the example of the proton pump being, in some way, undermined by the quark pump (5D22). And he refers to the experimental difficulties of the field, in a way which suggests that the quark pump would have replaced the proton pump because it dealt more satisfactorily with the experimental findings (5D25–7). The suggestion seems to be that the quark pump would succeed because it is experimentally better. It may be necessary to wait ten or fifteen years (5D20), not simply for one true version to be contingently displaced by another, but for the experimental inadequacies of the first version to be demonstrated and for a more experimentally sound theory to emerge. To some extent, then, these statements seem to be grounded in some kind of empiricist perspective. Yet, at the same time, the respondent's final statement in the first passage is carefully conditional and instead of ending with a phrase like the previous speakers' 'eventually we will know what's going on' in the biochemical realm, he concludes with the more sociological concept of beliefs *coming to be accepted* as proven (5D27). Moreover, he never says *explicitly* that the quark pump was scientifically better than the proton pump. All we know for certain is that 'somebody else might come up' who rejects the proton in favour of the quark pump. The grounds for this rejection remain implicit.

Our suggestion, then, is that both empiricist and contingent elements are to be found in the first passage and, indeed, that this speaker, like those quoted above, tries to use a notion of temporal development to resolve the interpretative problems which are thereby engendered. However, as we have already shown, it is not easy to use the notion of temporality to provide strong justification for one's action and beliefs within a contingent framework. Only by disregarding the contingent features of the first passage, can one treat it as an effective TWOD. And if one reads the first passage in the contingent terms of the second passage, the potential TWOD ceases to function as such.

The question, therefore, arises of why contingent as well as empiricist elements are found in the first passage and why the powerful contingent statement of the second passage is devised. Clearly it cannot be due to a pronounced personal inclination towards the contingent position. For, in addition to Cookson's interpretative ambivalence in the first passage quoted here, he uses an empiricist repertoire in other parts of the transcript to stress that his theoretical views follow unproblematically from the experimental evidence. We suggest that the contingent repertoire comes to predominate in these passages as the speaker engages in the unusual practice of acknowledging his own errors and making these errors scientifically legitimate.

In sentences 1 to 8, Cookson is answering a question about why he had written in a review article seven years earlier that Spencer's theory was impossible to disprove. One interpretative problem facing him in this section is that in the course of the interview he had already stressed that his present enthusiastic advocacy of Spencer's theory had been brought about 'strictly by the evidence'. His task in this passage, therefore, is to explain a criticism he had previously made of a theory which he now says is unambiguously warranted by experiment. He begins by saying that at that time he felt that Spencer was wrong and he goes on to try to specify what he had meant in the article. He clearly distances himself from his earlier views: 'I felt at that time', 'I was trying to say', 'I wasn't really', 'It was taking the position that'. He does say that 'it was a very personal kind of statement' (5D4). But this phrase works to separate the personal statement of the chemiosmotic unbeliever of seven years ago from the present speaker. It also suggests, however, that questions of proof and disproof can be dealt with in variable ways, depending on personal inclination. In the following sentence (5D5), he tries once more to state what it was he had been trying to express in the article. Then in sentence 6 he appears to abandon the attempt to recapture the logic of his previous position. He stops in mid-phrase and re-states his present opinion, that is, Spencer turned out to be right and the speaker to be wrong. This recognition of his error and of his mistaken attempts to undermine Spencer's theory leads him to formulate a rule of conduct: namely, the chances of being right are low and, therefore, one should not become too committed to one's present views. It follows from this rule that the recognition and abandonment of one's own erroneous views is a sign of scientific merit.

By the end of sentence 8, Cookson has made his own prior error seem allowable, if not commendable, by drawing attention to how hard it is to be right. He still seems to be organising his discourse in empiricist terms.

He has in no way revised his previous claim that Spencer had been shown to be right strictly by the experimental evidence. Nevertheless, getting it right has been portrayed as an unpredictable business and attention has been drawn to the fact that one can never be quite sure when one has done so. At this point, the interviewer asks why it is so difficult to be sure that one is right and suggests that it may have something to do with the nature of the experimental evidence (5D9). The respondent takes up this suggestion enthusiastically, mentions some of the factors which make the evidence ambiguous and then searches for an example. The example he chooses, that of his own work on alpha beta, is another instance where he now states that his previous scientific interpretation was wrong in certain respects. He describes his initial negative response when Stephenson first challenged this work (5D17). The speaker explains that he now accepts that Stephenson has successfully questioned his findings by 'doing something we didn't think of' (5D18). Then he proceeds to develop the notion of proof as a time phenomenon. The subsequent discussion of proof, with its combination of empiricist and contingent elements, serves, like the previous account of interpretative uncertainty, to put the speaker's error in a reasonably favourable perspective. His argument shows that it is not just his own theories which are overthrown; even Spencer's theory may be shown to be wrong and abandoned in time. The most that anybody can do, he concludes, is to make a lot of observations, provide a consistent interpretation and hope that you are not shown to be wrong (5D27).

In sentences 16 to 27, then, Cookson reflects upon his second error, develops the notion of experimental uncertainty, moves his exposition of the nature of scientific proof in a contingent direction, and provides moderate justification for his own mistakes by making them appear expectable. Both sentences 16 to 27 and 1 to 8 are organised, in part, so that the speaker's errors are made to appear normal and, thereby, legitimate. There then follows in the transcript an unquoted discussion of related topics which eventually gets round to general talk about the evaluation of experimental findings. Suddenly Cookson returns to the alpha beta episode and formulates clearly what he takes to be its epistemological implications (sentences 28 to 34). Truth is now viewed as that which most people are willing to accept at a particular point in time. In this passage, the respondent achieves a stronger, and less ambiguous, justification than before for his theoretical error. His views on alpha beta become as true for their time as Stephenson's or Spencer's for theirs.

With these epistemological assertions Cookson concludes his conversational turn (5D34). He has finally produced a comprehensive, contingent version of scientific belief which furnishes potent legitimation for his having advocated beliefs which he now accepts are wrong. Because this is such a forceful formulation, the kind of temporal concept which Cookson had used previously, and which is essential to the empiricist TWOD, is unnecessary to make it effective as a means of justification. The idea of movement towards a final or a more adequate stage of scientific proof is redundant when every stage has more or less equally uncertain scientific merit. That which is now widely regarded as wrong becomes as scientifically legitimate, for its advocates, as that which is now taken to be correct.

It appears from this case that the fully-fledged TWOD, and the associated empiricist perspective, is likely to be used in those typical situations where the speaker is reconstructing events in a way which directly displays the correctness of his current scientific views. The TWOD is less likely to be appropriate in cases where the speaker is engaged in making his errors understandable and scientifically acceptable. In the latter situation, it seems that a more effective reconstruction of the speaker's actions can be achieved by setting them within a strongly contingent portrayal of scientific action and belief.

Conclusions

In this chapter we have explored one implication of our previous analyses of scientific discourse in terms of the empiricist and contingent repertoires. Our discussion of these two repertoires implied that scientists will tend to have some difficulty in using them together in particular passages of conversation. We have looked closely at four passages in order to observe whether such difficulty is evident and whether any regular interpretative mechanism is employed as a means of combining the repertoires in a way which resolves potential contradictions. We have shown that the 'truth will out device' is used in this way in three of these passages. It is clear that the same pattern occurs in all but one of our collection of ten instances. Although our collection of TWODs is small, its use by nine of our 34 respondents suggests that it is reccurent in scientists' discourse and that it has some interpretative usefulness.

We have shown that the TWOD occurs in situations where the two repertoires are used together in relation to the same sets of actions and where the potential incompatibility of these two perspectives is signalled by the speaker's tendency to hesitate, revise his claims and offer apparently contradictory statments. The TWOD is able to reconcile the potential contradiction between the empiricist and contingent perspectives because it assumes a clear distinction between empiricist and contingent influences on scientific belief and then separates them temporally. Experimental evidence is depicted as becoming increasingly clear and conclusive over time and as enabling scientists to recognise, discount, and eventually eliminate the influence of contingent factors. Thus the TWOD is a device for re-establishing an interpretative divorce between the two conflicting repertoires when they have been combined in one conversational passage. More generally, it is an interpretative resource which enables speakers to resolve the potential inconsistencies that periodically arise as they generate diverse accounts of their social world.

The main effect of the TWOD is to restore the primacy of the empiricist repertoire. As we have seen above, this dominant repertoire in scientific discourse enables speakers to treat their own views as requiring no justification beyond reference to 'experimental facts', which are treated as coincident with the real world. When this perspective is adopted in ordinary talk, a speaker's scientific assertions become indistinguishable from the empirical world under investigation.[2] The TWOD enables speakers to return to this relatively secure interpretative domain after having ventured into the more difficult realm of contingency, within which scientific belief and the speaker's views become uncertain and difficult to justify with any finality.

Why, then, do scientists ever depart from the empiricist repertoire in the course of informal interviews? One reason for this is that there are social phenomena in science which are informally given meaning by use of the contingent repertoire and which can only be described and, indeed, can only be said to exist in so far as the contingent repertoire is used. For instance, actions regularly occur in science which can be said to involve scientific disagreement, but which are typically described as personal antagonisms, intellectual rivalries, competitions for status, clashes between strong personalities, and so on. It is clear that in much informal talk among scientists such things are discussed, are taken to be an essential part of science and are depicted as influencing the course of scientific development. As we have seen in this chapter, scientists sometimes hesitate before employing the repertoire through which such phenomena are constituted. They tend to preface their remarks by phrases like 'I'm not sure *I* want to talk about . . .' or 'If I am to be entirely honest, I must . . .' Nevertheless, despite scientists' preference for the empiricist mode of discourse, it is inevitable that references to this aspect of science will be made, and its constitutive repertoire employed, as scientists draw on their past interpretative activities in order to make sense of their world for the interviewer. Indeed, although speakers sometimes hesitate before drawing attention to the realm of contingency, we have seen above that at the same time they tend to emphasise that its importance in science cannot be denied. Consequently, it is to be expected that both repertoires will be used by scientists informally and that interpretative problems involving the two repertoires will arise which require resolution by means of the TWOD or by means of similar, as yet unidentified, devices.

The TWOD is a very effective reconciliation mechanism. Because the speaker's own scientific views identify the boundary between contingent and non-contingent actions and beliefs, the speaker experiences no difficulty in showing that, as beliefs which he takes to be correct come to be accepted, so contingent factors drop away. In addition, the duration of the process whereby correct belief is established is open to variation, depending on the particular circumstances specified by the speaker and on his ability or inability to show that his views are coming to prevail. The TWOD is so constructed that any scientist who claims that his views are correct, whether or not those views have yet been widely accepted, can reconcile the two repertoires in favour of the empiricist perspective and, thereby, gain for his own views the powerful justificatory support of that perspective.

The flexibility of the TWOD is, however, not indefinite. It is difficult to formulate a convincing TWOD in conjunction with a strongly contingent account of scientific knowledge. It is also difficult to use the TWOD to account for acknowledged errors. There is, therefore, a strong interpretative link between claiming correct belief, justifying that belief through the empiricist repertoire and using the TWOD to re-establish that repertoire. Conversely, acknowledging incorrect belief can be given greater legitimacy by use of the contingent repertoire which, in its strongest form, makes resort to a 'truth will out device' unnecessary.

These are our tentative conclusions in this chapter. Although they are based on analysis of a small number of cases, they are consistent with, and receive support from, the content of our previous chapters. Nevertheless, because we are claiming to have observed a recurrent phenomenon which is linked to the existence of interpretative perspectives adopted throughout science, the analysis should not stop here. Further close examination is required of discourse produced by other scientists in similar and in different interpretative contexts. The kind of fine-grained exploration attempted in this chapter exemplifies one direction in which our kind of analysis can be extended, that is, towards an increasingly detailed 'natural history' of scientists' discursive practices. But this is by no means the only possible line of analytical development. In the next chapter, we will explore a different avenue of investigation opened up by discourse analysis.

6

....

Constructing and deconstructing consensus

In the chapters above, we have begun to develop an analysis of scientists' accounting procedures; that is, an analysis of the procedures exhibited in various kinds of scientific discourse. We have stressed that even scientists' written technical discourse involves the representation of participants' actions and beliefs; and we have illustrated how scientists' actions and beliefs, like those of other social actors, can be characterised in numerous different ways by different actors and by the same actors in different interpretative situations. We have emphasised that scientists employ certain stable interpretative forms and repertoires, but that these recurrent interpretative resources are used with great flexibility to generate radically different accounts of social phenomena. We have found that participants' accounts are so contingent and variable that it is impossible to produce conventional sociological interpretations which are derived from these accounts in a satisfactory manner. We have suggested, accordingly, that instead of attempting to use participants' accounts as the basis for definitive analysts' versions of scientists' actions and beliefs, we should concentrate on identifying the principles in terms of which scientists' own accounts of action and belief are organised.

In the present chapter, we will extend this approach to deal with the issue of cognitive consensus in science. At first sight, this topic may seem to be difficult to reconcile with our stress on the variability of scientists' social accounting. For, by definition, consensus can only be said to exist when there is considerable agreement amongst the participants. We will show, nevertheless, that consensus is best conceived as a contextually variable aspect of scientists' discourse about action and belief. In so doing, we will try to allay any doubts that the reader may have about the possibility of using our form of analysis to deal with the *collective* phenomena with which sociology has been customarily concerned. For cognitive consensus in science is this kind of collective phenomenon *par excellence*. Our examination of consensus in this chapter is intended to show that our form of analysis does not stop at the description of participants' interpretative methods, but can also reveal how participants use their interpretative resources to construct the realm of collective phenomena. Thus, we will begin to indicate in this chapter how discourse analysis could be used to revitalise an issue of longstanding sociological significance, the analysis of aggregate phenomena.[1]

Spencer's consensus diagram and its readings

In 1974, Spencer was awarded an important scientific medal for his work on bioenergetics. On receiving the medal, he gave an honorary lecture to the Biochemical Society. This lecture was subsequently published in a journal of biochemistry.[2] It is clear from our interview transcripts that all our respondents were familiar with its content; and in particular with the diagram reproduced here as 6B. The lecture is divided into two main parts. The first part presents the four 'basic postulates' of the chemiosmotic hypothesis, describes in chemiosmotic terms some of the detailed processes of oxidative and photosynthetic phosphorylation, and links the chemiosmotic 'rationale' to the wider biochemical literature. In the second part of the paper, Spencer reviews some of the evidence which has led to the 'establishment of the four fundamental postulates of the chemiosmotic hypothesis as experimental facts'.

In between the two main sections there is a short 'historical comment on trends of opinion' concerning the chemiosmotic hypothesis, accompanied by a graph. This graph and most of the written text of his historical comment are reproduced below. As in the last chapter, we have numbered the sentences in this passage for ease of reference and we will do so elsewhere in this chapter wherever necessary.

> 6A
>
> 1 When I first began to develop and advocate this chemiosmotic view of oxidative and photosynthetic phosphorylation in the early 1960s, the four fundamental postulates were almost entirely hypothetical, many of my most distinguished and respected colleagues, such as Tippett, Holst, Bridge, Brian, Parry, Dowland, Arnold, Bull and Purcell were persuasive supporters of coupling through energy-rich chemical intermediates and coupling factors; 2 and there did not, perhaps, seem to be much chance that the chemiosmotic hypothesis would survive the destructive experimental testing to which, we were all agreed, it should be subjected. 3 Nevertheless, it was incidentally my hope that the chemiosmotic view would survive, because, if it did, there was . . . the chance that it might end the debilitating lack of agreement between the experts in oxidative phosphorylation and related energy transductions by providing the foundation for a generally acceptable conceptual framework . . . 4 As it turned out, the research sparked off by the chemiosmotic hypothesis in

many laboratories, including my own, produced much experimental evidence in support of the four fundamental postulates, which are now widely recognised as being experimentally established facts. 5 However, although there is now a relatively widespread acceptance of the chemiosmotic rationale in the field of photosynthetic phosphorylation, where there is a strong biophysical tradition, there has been more resistance in the field of oxidative phosphorylation. 6 Several of the more eminent authorities in the field of oxidative phosphorylation are reluctant to agree that coupling between the proton-translocating respiratory chain system and the proton-translocating ATPase system, plugged through the coupling membrane, is due to the proton current circulating between and vectorially through them. 7 They have preferred to believe, in keeping with the traditionally scalar origins of their conception of metabolism, that coupling is achieved by some unidentified energy-rich intermediates or by some direct interactions between components of the respiratory chain and reversible ATPase systems. 8 In [the figure below] I have plotted an assessment of the attitudes of some of the principal protagonists, based on that given by Cranmer in his excellent scrutiny of the chemiosmotic hypothesis (1969), and extended over the period from 1961 to 1973. 9 Obviously, different research workers judge the same experimental knowledge differently, and opinions change as time allows improvement of comprehension and accumulation of knowledge. 10 Looking at the trend shown by this diagram, and bearing in mind the natural and inevitable predilections of the different protagonists, it does seem likely that the validity and usefulness of the chemiosmotic rationale in the field of oxidative phosphorylation and related energy transductions will be generally recognised in due course. 11 It is understandable, however, that some of the eminent biochemists, who have long championed the more traditional biochemical views of the coupling mechanism, have not found it easy or agreeable to acquire a taste for the relatively biophysical disciplines of membrane transport and vectorial metabolism that were not originally of their own choosing.

Diagram 6B is presented by Spencer in the text of his lecture as a straightforward description of the changing pattern of support for chemiosmosis. It is taken as documenting how the opinions of relevant specialists with respect to chemiosmosis have actually altered in the past and it is used (6A10) to suggest what is likely to happen in the future. The written text takes for granted the accuracy of the diagram and offers an interpretation of why the trend should have taken this form. The growing consensus about the scientific merits of chemiosmosis which is displayed in the diagram is attributed to the gradual accumulation of experimental evidence (6A4) and to the slow improvement of comprehension (6A9). The absence of a complete consensus is linked, in accordance with the

6B

Trend of support for chemiosmotic rationale

Research worker	Date			
	1961	1965	1969	1973
Holst				
Brian				
Parry				
Dowland				
Purcell				
Bridge				
Warlock				
Arnold				
Tippett				
Byrd				
Bull				
Tallis				
Gibbons				
Boyce				
Handel				
Elgar				
Britten				
Spencer				

☐ Supporting chemical or direct-interaction coupling.

▨ Supporting chemiosmotic rationale.

asymmetrical structure of accounting for error, to reluctance on the part of certain eminent scientists to adopt a theory which they themselves had not originated and which they found rather difficult to understand (6A5–7 and 11).

In the course of our interviews carried out in 1979, several of our respondents mentioned this diagram or similar diagrams which Spencer seems to have produced at various times. We asked these and our other respondents for their comments on the diagram. Some of them accepted it as an accurate, literal description of the growth of cognitive consensus in the field.

6C

Interviewer: I wondered if you'd like to comment on how accurate you think that is.

Shaw: Oh, I think its pretty accurate, just from the people I know . . . [Shaw, 48]

6D

Interviewer: You've undoubtedly seen that before. Can you regard that as an accurate account, as far as you can judge?

Miller: Yes, I think so. Not bad. I'd never thought about the whole chart before. I've just looked at it and laughed. I'm not certain about the dates, but the *sequence* is correct. [Miller, 20]

6E

Interviewer: Do you think that is a fair representation?

Waters: I think so, yes . . . I think in a stepwise fashion more and more people are being convinced. [Waters, 7]

Many of our respondents, however, questioned Spencer's diagram and, in doing so, focused on three interpretative problems which Spencer had apparently been able to resolve.

6F

The people at the top were in the field at the beginning . . . My name is deliberately left off . . . Handel himself is not in the field. Tallis is not in the field. Byrd is not in the field. Bull certainly *was* in the field . . . I don't think they had the foggiest interest as to what was the mechanism . . . I would miss out everybody except Arnold, Bridge, Purcell, Dowland, Parry, Brian and Holst. [Jennings, 10–11]

6G

At this date, [Spencer] switched me from being a non-believer to being partially convinced. I think he is half right. Which is a difference. [Warlock, 12–13]

The implication is we can't make up our mind. I don't think we are any more uncertain than anybody else. It is just that we're as I have said, not that I am half convinced, but I am fully convinced he is half right. [Warlock, 52]

6H

People in this field . . . will say something that is really contradictory to the Spencer hypothesis and they will still declare that they are Spencerians . . . Spencer himself has different versions and that's I think one of the confusing things. That when somebody says that Spencer is right or chemiosmosis is right, you would really have to nail down exactly what is meant . . . You can't decide who is a Spencerian and who is not, if you don't first of all define exactly what the doctrine is. [Hinton, 10–11]

Each of these speakers objects to Spencer's diagram and states that it is misleading in a particular respect. Jennings (6F) maintains that Spencer has incorrectly defined the membership of the field. He points out that he has been omitted and various scientists, he claims, have been wrongly included. Although he does not specify clearly what criteria qualify biochemists for proper membership of the field, he seems to suggest that some kind of active interest in formulating a *mechanism* of oxidative phosphorylation is a necessary prerequisite. It follows from this speaker's criticisms that, if the membership of the field had been 'properly identified', the shape of the curve of consensus would have been quite different.

The next speaker, in quotation 6G, makes it clear that Spencer's consensus diagram requires scientific belief to be correctly attributed to the members of the field. This speaker asserts that he, and others on the list, have been wrongly categorised. He is not, he says, half persuaded that the whole of chemiosmosis is right. He is, rather, certain that there is an element of truth in the hypothesis; and equally certain that 'most of the details are nonsense' [Warlock, 12]. This speaker maintains, then, that his beliefs, and those of some of his colleagues, do not fit the categories provided by the diagram. Clearly, the accuracy of the diagram and of the accompanying text depends on Spencer's ability to recognise correctly and to cope with the complexity of other participants' scientific views.

The third speaker develops further the issue of how scientific belief is to be attributed. He claims that, for him at least, the meaning of the term 'chemiosmotic hypothesis' is unclear and that, in his opinion, other scientists appear to interpret the hypothesis in various ways, some of which are incompatible with Spencer's own position (which is itself said to be variable). As another respondent put it: 'If you read papers by people who are, in a sense, saying that they believe in the theory, they will often say "my interpretation of the Spencer theory etc., etc.". Which is perhaps not Spencer's theory at all. What sort of theory is it, that each person has to put their own gloss on it?' [Sephton, 15]. These scientists question the value of the consensus diagram on the grounds that, although over time more scientists may have come to profess acceptance of something which they call 'the chemiosmotic hypothesis', these scientists are hiding important differences of scientific opinion behind a superficial terminological agreement. If the chemiosmotic hypothesis means something different for each participant, then its increasingly widespread verbal endorsement in no way indicates that there is growing uniformity of scientific belief within the network.

The interview quotations above show that, although Spencer's consensus diagram *can* be read as a simple, literal description of what has happened in the field (6C to 6E), it can also be read as being seriously misleading (6F to 6H). The last three quotations illustrate the grounds on which typical objections to the diagram were based in our interview material. We suggest, however, that these latter quotations have a wider significance than this. For it appears that they address three basic interpretative issues which have to be resolved in *any* claim to describe the state of cognitive consensus in a scientific field. First, in claiming consensus it is implied that the speaker has identified all the relevant members of the field, that is, all those competent scientists who must be considered as working within the area of investigation in question. Secondly, it is implied that the speaker can attribute scientific belief correctly to each individual scientist. Thirdly, it is implied that the cognitive content of the consensus can be specified accurately and shown to coincide with the views of all those who are said to belong to it.

Analysts' and participants' consensus claims

These interpretative issues are not only faced by scientists as they construct their accounts of consensus, but also by any sociologist who attempts to formulate claims about scientific consensus in general or about the degree or nature of consensus within specific research networks. Consider, for example, Ziman's assertion that scientific 'facts and theories must survive a period of critical study and testing by other competent and disinterested individuals, and must have been found so persuasive that they are almost universally accepted. The objective of Science . . . is a *consensus* of rational opinion over the widest possible field.'[3] In order to verify such a claim, one would have to review all those specialties included under the rubric 'science', identify the competent specialists in each case, and either formulate or locate formulations of the bodies of knowledge to which each of those groups of specialists are committed.

When one examines the secondary literature on science, however, it is evident that, despite the frequent assertion that science is unique in its attainment of cognitive consensus, there are no studies available which delineate in detail the nature and extent of consensus within any particular research network. Moreover, the few studies in which the topic of scientific consensus has been empirically explored show that *the analyst is ultimately dependent for his conclusions on the interpretative work carried out by participants.* Although scientists produce various texts, such as review articles, textbooks and research papers, which, it is suggested, can be used as unobtrusive measures of consensus, scientists' actual beliefs can never be inferred directly from these literary products alone. We have illustrated this in the preceding section.

The usual sociologists' solution to this problem is to supplement or replace such documentary material with direct questioning of scientists. Such an approach is illustrated in one of the few systematic empirical studies of scientific consensus, in which respondents were asked the following question: 'If you consider the relevant literature on this *speciality* field, how much *agreement* is there with respect to the *theoretical* approaches which ought to be applied. . . those techniques and methods that can be considered as generally accepted . . . the *results* which so far can be considered as generally *accepted?*'[4] In this study by Knorr, 'degree of consensus' becomes equivalent to 'percentage of respondents claiming a high degree of consensus'.[5] The analyst presents respondents with a question which requires them to judge what is the relevant literature, what are their colleagues' scientific beliefs, what is agreed, and who is involved in the consensus; and then the analyst offers the aggregated responses as a measure of consensus.

Clearly, such an approach means that the empirical findings are generated out of scientists' own answers to the underlying questions of membership, attribution of belief and cognitive content. Scientists' reports of consensus could be used satisfactorily in this way, only if the reports could be taken as simple, literal descriptions of a state of scientific belief. But such an assumption is difficult to sustain. A similar assessment of the degree of consensus or the use of a similar label to identify the content of a supposed consensus may mean different things for different speakers. Furthermore, as we will see below, not only is it possible in a specific case for different scientists to give quite different views of the state of cognitive consensus, but each individual scientist, in different interpretative contexts, can furnish quite different accounts of the degree and nature of consensus in a field.

Such interpretative diversity occurs, we suggest, because the meanings of the underlying issues of membership, individual belief and cognitive content are themselves contextually defined and contextually variable. In view of the analytical position presented in our opening chapter and in view of the substantive analyses developed in subsequent chapters, this is hardly surprising. The particular significance of such interpretative diversity here is that it reveals how the data customarily used by sociologists as indicators of consensus are context-linked interpretative products, arising out of participants' solutions to the three underlying issues. Thus we begin to see that to adopt the traditional approach to the analysis of this collective phenomenon, as exemplified above in the writings of Ziman and Knorr, is to do no more than to present the aggregate results of scientists' own 'sociological theorising' about consensus as if it were itself the phenomenon of scientific consensus.

One implication of this argument is that we need to know more about the way that scientists carry out their 'sociological theorising'. We need to investigate how participants create the appearance of shared belief and how they construct their solutions to the three underlying issues identified above. In the next section we will take a step in this direction by returning to Spencer's diagram and its accompanying text and by examining how Spencer deals with these issues. We will show that participants are able to furnish plausible denials as well as assertions of growing consensus, depending on the interpretative procedures they use to deal with the issues of competent membership, attribution of belief and the content of theoretical categories. The discussion which follows will begin to show how the collective phenomenon of scientific consensus becomes amenable to fruitful empirical analysis only when it is conceived, not as a social fact *sui generis*, but as a contingent product of participants' variable interpretative procedures.

Defining the field and identifying its members

In 6A, Spencer states that his diagram plots the attitudes of some of the principal protagonists in the field and that it is based on a review article published in 1969 by Cranmer (6A8). Cranmer describes his review as covering research on oxidative and on photosynthetic phosphorylation; and Spencer's diagram includes scientists specialising in both these areas. It seems, therefore, that 'the field' to which Spencer's consensus diagram refers can be seen as being composed of at least these two smaller areas of research. Spencer himself draws attention to this division, when he claims that there has been more resistance to chemiosmosis in '*the field* of oxidative phosphorylation' than in '*the field* of photosynthetic phosphorylation' (6A5). Furthermore, in interviews, respondents tended to describe themselves as specialists in one or other of these areas and, although often emphasising that the oxidative and photosynthetic systems are scientifically similar, they stressed that in many respects these sub-fields were intellectually and socially distinct.

On some occasions, then, our respondents treated oxidative and photosynthetic phosphorylation as parts of a single research area. On other occasions, these two realms of study were treated as being relatively separate. However, the shape and significance of Spencer's consensus curve depends crucially on his treating the two potentially separate fields as one in composing *this* diagram. For instance, the top six scientists named on the diagram, who are classified as remaining non-chemiosmotic throughout the whole period, are regularly identified as specialists in oxidative phosphorylation. In contrast, those at the bottom of the diagram, who are shown as quickly coming to support chemiosmosis, are said to be preponderantly involved in photosynthetic phosphorylation. Only Spencer and one other among the bottom five names are not defined in Spencer's own text and elsewhere as being mainly concerned with photosynthetic phosphorylation. Thus, it is only by treating the two 'fields' as one scientific entity that Spencer is able to produce his smooth curve of increasing cognitive consensus. If he had adopted in his diagram the distinction between the fields which he employs in his written text, he would have presented two distinct curves. Assuming the same personnel, one of these would presumably have shown very rapid conversion to chemiosmosis in the area of photosynthetic research. The other would presumably have depicted oxidative phosphorylation as remaining largely unaffected by chemiosmosis up to 1973. If this had been done, Spencer's claim, that merely looking at the trend makes it seem likely that his theory of 'oxidative phosphorylation and related energy transductions' will be generally accepted in due course (6A10), would have been seriously weakened, if not completely undermined.

This is by no means the only problematic aspect of Spencer's choice of persons to represent the changing trend of scientific opinion. The bibliography to Cranmer's review, on which Spencer's diagram is said to be based, contains 115 different first authors. Spencer's diagram plots the views of only 18 scientists. Spencer gives no clear indication of how the 18 were selected from the larger pool. He does suggest that he has focused on 'some of the major protagonists'. But no explanation is given of what exactly is meant by 'major protagonist' and no clear rules of inclusion/exclusion are presented which would enable us to arrive at the particular list of scientists used by Spencer.

Spencer seems to have dealt differently with the top and bottom of the diagram in another respect. The top nine names, largely non-chemiosmotic, are all leading scientists and heads of laboratories who were well established in 'the field' before 1961, the first date on the diagram. In the bottom half, however, we have three scientists who were not in the field in 1961 and one of these, Handel, entered the field as a student of an older scientist also included in the bottom half of the diagram. It is by no means clear why these chemiosmotically inclined new entrants are included in the diagram; nor why some students of those scientists resisting chemiosmosis are not also included. Once more, the procedures for identifying the membership of the group on which Spencer's claim is founded seem equivocal and contingent. Yet, in so far as one reduces or expands Spencer's list, one necessarily alters the characterisation of the pattern of consensus embodied in his diagram.

One further possibility that has to be considered is that Spencer is simply adopting those explicit judgements of support for chemiosmosis which are contained in the opening paragraph of Cranmer's review. But, although Cranmer's list overlaps with Spencer's, they are not identical. Spencer leaves out one of Cranmer's critics of chemiosmosis and two of his neutrals and introduces three opponents of chemiosmosis, as well as two scientists who are shown as becoming fully converted by 1973 and one who is shown as half convinced by that date. Another major difference between Cranmer and Spencer is that the former's listing of those for and against chemiosmosis is relatively casual and does not claim to be representative, while Spencer, in contrast, uses his similar sample to depict formally the 'trend of support for the chemiosmotic rationale'. The biggest difference, however, between Cranmer and Spencer is that whereas the former offers a sketch of the situation at a single point in time, the latter extends his diagram to cover the period 1961–73. In this respect Spencer goes far beyond Cranmer and it is clear that Spencer's text is 'based' on that of Cranmer only in the loosest sense.

We suggested in the previous section that traditional sociological analysis of the 'collective phenomenon' of scientific consensus was dependent on prior interpretative work carried out by participants in relation to three underlying issues, one of which was group membership. We have now seen that Spencer's portrayal of the pattern of consensus is intimately linked to the way in which he selects particular scientists to represent the field; that various quite different lists could have been compiled with equal plausibility; that the choice of different scientists, given unchanged attribution of scientific opinion to each individual, would have produced a significantly different picture of the changing pattern of cognitive consensus; and that Spencer's procedures for making his selection remain unspecified.

These observations suggest that Spencer did not first resolve the issue of membership through the application of clear-cut, objective criteria and then find that he had an upward consensus curve. Rather, it seems that the task of constructing such a curve informed all his judgements about membership. In other words, Spencer's consensus curve is a creative display of his own interpretation of the changing nature of consensus in 'ox phos', which is achieved partly through tacit judgements about group membership and which achieves an appearance of facticity partly by treating the issue of membership as unproblematic.

Spencer, of course, never claims that the scientists named in 6A constitute a definitive or comprehensive membership list. Nevertheless, the upward consensus curve, which is utterly dependent on Spencer's choice of personnel, *is* treated in Spencer's written text as an accurate depiction of the actual movement of scientific opinion within a genuine field of biochemical research. Although the diagram appears analytically to be highly contingent, it is presented in Spencer's text as a literal description of an underlying social reality. For sociological purposes, however, it is clear that we cannot regard Spencer's diagram as documenting a consensus which exists independently of the interpretative work embodied in that diagram. The appearance of consensus exists only through this interpretative work; which has to be conceived analytically, not as a description of a collective phenomenon 'out there' in the field of bioenergetics, but as an interpretative accomplishment achieved by Spencer on the occasion of his honorary lecture and available as an interpretative resource to others as they create *their* versions of the history of 'the field'.

In the previous section we saw that, despite the apparent contingency of Spencer's consensus claim, many scientists were able to treat it as an accurate description and indeed, like Spencer himself, as merely stating the obvious. For such scientists Spencer's diagram, whatever its possible inadequacies in detail, is treated as documenting a real, underlying pattern to be observed in their field. Spencer's diagram is taken simply as displaying 'what everyone knows'.

> 6J
> *Interviewer:* I wondered . . . whether you thought his overall trends were accurate.
> *Perry:* [*Looking at 6B*] I would think so. Has anybody objected to it? I wouldn't expect it. I think I can tell you who would object to it probably . . . I don't think that's important. I think that basically it is accurate.
> [Perry, 12]

When scientists respond in this way and express agreement with Spencer, they never raise questions about his interpretative procedures. Like the speaker in 6J, they sometimes recognise interpretative problems, but they dismiss these as minor imperfections which in no way detract from the overall accuracy of Spencer's consensus claim. When respondents deny Spencer's claim, however, they can trade upon underlying interpretative issues as grounds for wholesale rejection of Spencer's account. In so doing, speakers can focus on the issue of membership of the field, as in quotation 6F. But more frequently they focus on the issues of the content of the supposed consensus and the attribution of belief to individual scientists.

Attributing scientific belief

In the review articles by Spencer and Cranmer, the views of individual scientists with respect to oxidative phosphorylation are identified only by means of allocation to categories such as 'supporting chemiosmosis', 'supporting chemical coupling', or 'adopting a notably neutral attitude'. The allocation of individuals to these categories is either undocumented or is warranted by references to one or a few of their published papers. The attribution of scientific belief is treated as unproblematic, in so far as no reservations are expressed in either text about its accuracy.

The method of attributing belief adopted by the authors of these reviews seems to involve several interpretative procedures. First, attributions are made as if each scientist, at any particular point in time, has a specifiable scientific view with respect to a range of biochemical phenomena. Secondly, it is taken for granted that each scientist's position can be accurately identified by other scientists from his research publications. Thirdly, it seems to be assumed that differences between individuals' beliefs are relatively unimportant and that the similarities between their views coincide neatly with the major hypotheses in the field. These general procedures appear to provide part of the interpretative resources which participants use to construct accounts of cognitive consensus.

Although these procedures are regularly employed when participants' interpretative work entails the unproblematic attribution of belief, they can be explicitly abandoned in other interpretative situations. Consider the following exchange.

> **6K**
> *Interviewer:* Did Waters fully understand the chemiosmotic theory?
> *Barton:* Not when I got there, no . . . he probably would have liked the chemical theory to have come through, after all. Because that's what he'd been personally committed to. But he was certainly absolutely objective and well capable of recognising the force of any other theory and I'm sure, by the time I left, he was in favour . . .
> *Interviewer:* I'm not very clear about Waters. Is there a point at which Waters comes to accept the chemiosmotic theory?
> *Barton:* I couldn't tell you that. You'd have to ask him that. I left in 1971 or so. [Barton, 14]

In this passage, the respondent initially has no qualms about attributing beliefs to his erstwhile supervisor, nor about categorising his views in terms of the two major theories. Waters is said not to have understood the chemiosmotic theory when Barton first arrived in his laboratory and he is described as having been personally committed to the other theory. Then Barton, as a way of illustrating Waters' scientific objectivity, proceeds to assert that he is sure that Waters was in favour of the chemiosmotic theory by the time he left the laboratory. In the first paragraph, therefore, the speaker adopts an approach to the attribution of belief which is similar to that used by the authors of our review articles and by those scientists who accept Spencer's diagram at face value. Yet subsequently, on the same page of transcript, he denies the very claim he has just made about Waters having come to favour chemiosmosis. And he does this, not by stating that he had made a mistake about Waters' specific scientific beliefs, but by maintaining that it was quite impossible for him to ascertain at all what Waters' scientific ideas were. Thus, having confidently attributed various scientific views to Waters (I'm sure . . . he was in favour'), the speaker questions the very possibility of making any such attribution ('You'd have to ask him that').

What are we, as analysts, to make of such apparently inconsistent assertions? We suggest that scientists' ability, on occasion, to repudiate their own attributions of scientific belief, in the course of conversation, makes it difficult for analysts to accept such claims as anything more than contingent formulations which are devised in accordance with variations in interpretative context. Thus Barton's claim that Waters had come to favour chemiosmosis can be seen as a way of re-establishing Waters' scientific objectivity, which had been put in question by Barton's prior statements about Waters' personal commitment to the chemical theory and Waters' failure to adopt, at least initially, the theory which Barton portrayed as firmly established by the empirical evidence. This example shows, then, that the interpretative procedures identified above, by means of which participants attribute scientific belief, can be variably implemented and it suggests that they are used by a given speaker only in certain kinds of interpretative situations.

If attributions of belief in interviews are treated as context-dependent interpretative accomplishments, there seems no reason to regard them differently when they occur in review articles. The relative absence of obvious contradictions among the claims advanced in particular written texts is probably due to the care with which such texts are prepared, to the use of a restricted interpretative repertoire and to the absence of that direct interaction with other actors which elicits variable responses in so many subtle ways. Of course, even in the transcripts of informal interviews, the kind of blatant and immediate retraction exemplified in quotation 6K occurs only rarely. Taken alone, therefore, this passage cannot justify our claim that the attribution of scientific belief is highly variable and socially contingent. However, the divergent attributions made by different participants provide much additional evidence of this phenomenon.

We have a great many instances where the beliefs of particular scientists appear to be characterised very differently by different speakers. This material is illustrated in quotations 6L and 6M. These two passages have been chosen because they both refer directly to Spencer's consensus diagram (6B) and because they both deal succinctly with a number of leading members of the field.

6L

[The consensus diagram] was partly just to indicate rather pleasantly and lightheartedly that perhaps Max Planck wasn't right and that this thing [the curve of support for chemiosmosis] might eventually get to the top in not too short a time, which as it happened is exactly what it did. Well, it hasn't quite got to the top. I don't suppose it ever will . . .

People like Brian, who was very much involved with the other theory, were working hard to make sure that the chemiosmotic hypothesis was accepted as a theory, for the time being . . . I would have said his attitude, Purcell's attitude, Holst's attitude, all the big people . . . the people involved are marvellously willing *in the long run*, or they have been in my experience, marvellously willing to be altruistic and to work for the beauty of our subject, in the end. [Spencer, 61 and 57]

6M

1 Many people have accepted it grudgingly, but many others – I think you will find that the people who were the most committed to other things haven't actually accepted it. 2 They have redefined their own theories. 3 But I mean, in fact, if you *now* take what they say and translate it into his terminology, they are actually saying the chemiosmotic hypothesis. 4 But most of them would not admit – I am sure that, looking down this list, that Holst hasn't; that, Brian is in a funny situation, he in many ways is perhaps the most objective of scientists, but he is I suppose a fence sitter. 5 Parry, well he's redefined everything, he's saying the Spencer thing but in very metaphysical terms. 6 Dowland I don't think has been converted at all. 7 Purcell certainly hasn't. 8 Bridge just somehow avoids it, he just talks about other things completely. 9 Arnold doesn't. Arnold is now more antagonistic than he ever was. 10 So if you take the top half of those people [on the diagram], they haven't really been converted. [Harding, 21–2]

In quotation 6L, Spencer brings the consensus curve up to date. He suggests in the first section that the curve has now (1979) reached the top, that is, that almost everybody accepts chemiosmosis. In the next section, he mentions three leading figures explicitly and he refers to 'all the big people' as having acted altruistically, that is, as having abandoned their previous incorrect views, as having ultimately admitted the scientific superiority of the chemiosmotic hypothesis and as having come actively to advocate its general acceptance. Thus the current views of virtually all the major researchers in the field are depicted as being basically similar and basically chemiosmotic.

In contrast, the speaker in 6M gives a much more complex account of the current situation as he rejects the idea of a strong chemiosmotic consensus. He allows far greater variation between individuals. He treats the identification of belief as more problematic. And he summarises the overall situation as one of continued resistance, rather than conversion, to chemiosmosis. Harding contends at the outset that, not only have people adopted chemiosmosis grudgingly, but that many of them have not actually accepted it at all (6M1). He then seems to make a distinction between what people *say* they believe and what they really do believe. He suggests that, if you translate or restate the scientific claims of some of these people into Spencer's terminology, their views appear to be identical with the chemiosmotic hypothesis (6M2–3). But, he suggests, most of them will not admit this (6M4). Thus Harding is proposing here that the identification of individuals' scientific views is by no means straight-forward and that it may sometimes involve a process of translation into terms rejected by the scientist himself (6M3).

At this point (6M4) the speaker, after a pause, refers to Spencer's list of names and produces brief summaries of their scientific views. He seems, in this sequence, to be describing what each of these scientists really believes and not merely what they say. Thus, Parry is 'saying the Spencer thing but in very metaphysical terms'. The others, however, and this includes all the 'big people' described by Spencer as accepting chemiosmosis, are depicted as not really believing at all in that theory (6M6–9). Moreover, Harding attempts to discriminate between different shades of scientific opinion among this collection of scientists. Some of them, he suggests, can be translated into chemiosmoticists (6M3 and 5). Some of them are indifferent (6M4). Some of them are actively opposed (6M7 and 9). And the views of others cannot be ascertained (6M8). Thus, whereas Spencer, in the positive consensus claim above (6L), treats the views of this group of scientists as similar, as essentially chemiosmotic and as illustrating the completion of the trend identified in the consensus diagram, Harding emphasises the current diversity of views, the lack of consensus with respect to chemiosmosis and the failure of Spencer's consensus curve to continue up to the present day.

In making this comparison between passages from Spencer's and Harding's interview transcripts, we have been able to illustrate how radically different scientific views can be attributed and frequently are attributed to given participants by different speakers; and to show that the identification of individual scientists' views by other participants is a complex and potentially variable interpretative achievement. We saw in the last quotation, as we saw in 6F to 6H, how scientists tend to make explicit and to undermine the kind of interpretative work involved in presenting a positive consensus claim as they warrant their own rejection of such a claim. Thus, as Harding deconstructs Spencer's account of consensus in 'ox phos', he comes to question whether scientists' views can be easily discerned in what they say or write. He also questions the assumption that each scientist has a single, coherent scientific position. In other words, he begins to repudiate some of the interpretative procedures identified above which are treated as unproblematic in Spencer's and others' positive consensus claims.

We have also seen that the attribution of individual belief is, at least sometimes, closely associated with the use of the names of specific theoretical positions. This feature of our data may be due, to some extent, to the fact that our interviewees tended to talk about others' views in connection with Spencer's diagram, which is organised around a simple division between the chemiosmotic theory and all other theories. Thus, there is no guarantee that their accounts of consensus in other everyday situations would closely resemble in this respect those forthcoming in the interviews. Nevertheless, nobody objected to Spencer's diagram on the grounds that its categories were inappropriate or unfamiliar. Moreover, every speaker identified numerous scientists who were chemiosmoticists and others who were, or had been, committed to the chemical theory. Indeed, these simple theoretical labels were used time and time again by our respondents to identify groups of participants whose views could be treated as scientifically equivalent or identical, and who therefore constituted a scientific consensus which, in the limiting case of 'chemiosmosis', was sometimes presented as virtually coinciding with the research network itself. The terms 'chemiosmoticist', 'chemiosmotic hypothesis', 'chemiosmotic theory', 'chemiosmosis', 'Spencer's hypothesis', and 'Spencerian theory' were especially pervasive in our interviews. Let us examine how participants interpreted and used these terms in relation to the topic of consensus.

The meanings of chemiosmosis

Let us begin this section by looking at the meaning given by Spencer to the term 'chemiosmotic hypothesis' in his Nobel lecture. This lecture took place five years after that in which Spencer presented the consensus diagram examined above and shortly before his interview with us. Although Spencer does not offer any graphical representation of changing opinion in the later lecture, he does give a verbal account of growing scientific consensus which brings that diagram up to date. One of the themes in this lecture is that research on biological energy transduction disproves Max Planck's famous dictum that new ideas are accepted only after their opponents die (see also 6L above). In his Nobel lecture, Spencer begins by stating that 'what began as the chemiosmotic hypothesis has now been acclaimed as the chemiosmotic theory'. This theory, he suggests, is designed to answer three elementary questions about respiratory chain systems and analogous photoredox systems: 'What is it?', 'What does it do?', 'How does it do it?' He continues, 'we can now answer the first two [questions] in general principle, and . . . considerable progress is being made in answering the third'. Thus 'with few dissenters, we have successfully reached a consensus in favour of the chemiosmotic theory'.

In the course of the lecture, a body of experimental evidence is reviewed and clear-cut answers provided to the initial questions. Although Spencer notes that there is still much to be understood about the details of the biochemical processes involved in energy transduction, he emphasises that chemiosmosis is an empirically concrete and experimentally validated series of propositions which describes in some detail the structure and functioning of respiratory and photoredox chain systems, which also explains the coupling between respiration and oxidative phosphorylation, and which extends beyond this limited range of phenomena to provide general principles applicable to other significantly different biochemical systems.

These points were repeated in our interview with Spencer, in which he referred once again to the existence of a 'pretty broad consensus' in favour of chemiosmosis. However, immediately after making this point, Spencer went on to point out that other scientists did not always fully understand what 'chemiosmosis' means. In other words, their versions of chemiosmotic theory differed from his.

> 6N
> Of course, people tend to take a very simplified view of a theory and say that *is* the theory. People have tended to say the chemiosmotic theory says that protons must go right out into the bulk [acqueous phase] and come back from the bulk. Well, it never said anything of the kind . . . I couldn't possibly fail to know that the surface conductance [at the outer surface of the membrane] is likely to be considerably higher than the bulk conductance. So I would never have been fool enough to say that they normally go right out. [Spencer, 70]

In this passage, Spencer identifies a particular 'misunderstanding' which some scientists have about chemiosmosis. In the passage which follows we find a scientist apparently exemplifying this misunderstanding.

> 6P
> 1 Spencer will just not consider anything about surface phenomena at all . . . 2 The interface between the membrane surface and the bulk just doesn't exist and you know, it damned well does! . . . 3 But I don't worry myself about it too much, while some other people will go around saying: 'Oh, the whole Spencer scheme is wrong because he's forgotten the interface.' 4 Of course, at the end of the day, it's relevant . . . 5 [But] I

don't think it matters that much . . . 6 What difference does it really make? 7 The concept is that you use the thermodynamic gradient of protons to make ATP. [Grant, 69]

In 6P, Grant describes Spencer's hypothesis in a way which we have just seen Spencer himself expressly repudiating; that is, he claims that chemiosmosis ignores the issue of surface conductance (6P1–2) and he proffers a highly simplified version of the concept of chemiosmosis (6P7). On these specific occasions, therefore, the meaning of chemiosmosis seems to differ for these two scientists. Grant refers to other scientists who treat Spencer's supposed failure to deal with surface conductance as integral to his theory and, therefore, as grounds for rejecting the theory (6P3). But Grant portrays himself as not agreeing with these scientists about the centrality of the phenomena of surface conductance to the chemiosmotic theory (6P5–6). Chemiosmosis, he suggests, can be stripped down to the basic notion that ATP is made by a gradient of protons (6P7). By redefining chemiosmotic theory in *this* way, Grant is able to separate himself from its critics and to present himself as a chemiosmoticist; as part of the consensus recognised by Spencer (6P3–6). Yet, in so doing, not only does Grant propose a version of chemiosmosis which differs in detail from Spencer's in relation to the specific issue of surface conductance, but he also promulgates precisely the kind of grossly simplified view of chemiosmotic theory which Spencer condemns at the beginning of quotation 6N. In other words, the appearance of scientific agreement between these two researchers is maintained only at the terminological level. It is accomplished by their both using the same theoretical label, namely, 'chemiosmosis', to refer to scientific interpretations which differ considerably in substance.

In our interviews, almost every scientist clearly operated with at least two versions of chemiosmotic theory. On the one hand, chemiosmosis was depicted as a theory dealing in some detail with the processes involved in oxidative and photosynthetic phosphorylation. At this level, the scope of the theory was similar to that covered in Spencer's Nobel Lecture. The content of the theory, however, and the degree to which it was taken to be experimentally validated, differed from one speaker to the next. There was little evidence of a uniform version at this level persisting from one speaker to another and much evidence of scientific disagreement. On the other hand, there was a highly simplified, basic version of chemiosmosis. This version was widely used by our respondents and can be seen as constituting, in some sense, a consensus. However, the scientific content of this basic version was minimal; so much so that it could be and frequently was endorsed by those who described themselves and were described by others as strongly opposed to chemiosmosis as well as by those who claimed to be fervent supporters of the theory.

Although Spencer and many others noted the widespread use of a simplified consensual version of chemiosmosis and although speakers frequently remarked on the diverse interpretations of chemiosmosis to be found in the research network as a whole, none of our respondents commented on their own use of two versions of chemiomosis. Rather, each respondent moved implicitly from one version to the other, as he talked about the theory and the degree of consensus, in accordance with the interpretative requirements of his discourse. In the following quotations, as in 6P, we can clearly see the speaker moving between a detailed and a basic version of chemiosmosis.

> **6Q**
> (a) 1 Even now there is this huge craziness [about stoichiometries]. 2 I think it is of relatively minor importance what the stoichiometry is . . . [although] it is an important question, because it has to do with the mechanism and generation of the electro-chemical gradient. 3 And of course Spencer's concept of loops is beautiful in its simplicity. 4 There's very little evidence for it and in my own case, I mean that business with the electron donors, finding inhibitors working at different steps, that still doesn't fit his loop theory and its hard to see how it could fit a loop theory . . . 5 People do not want Spencer to be right all the way. 6 They are willing to say 'OK well, the proton gradient has something to do with ox phos, but is it the only thing and is his stoichiometry right?' 7 That's nonsense. 8 There's *no* question in my mind that the overall theory is correct . . . 9 I can take every piece of data that I couldn't explain, except for the things that have to do with the loop and have to do with the generation of the electro-chemical gradient, I can take every piece of data and explain it . . . [Cookson, 15]
>
> (b) 10 [The evidence] certainly doesn't *favour* a loop mechanism. 11 Now I don't make a big thing out of that right now because, from *my* point of view which has always been the translocation of substrates, it's not that important. [Cookson, 20]
>
> (c) 12 I think that without question the overwhelming percentage of what Spencer has to say is right and I think there is very little that one can argue against that [Cookson, 21]
>
> (d) 13 I don't see the kind of dialogues taking place at meetings

that did take place 10 years ago. 14 There was a lot more
open discussion and debate 10 years ago than there is now . . .
15 One of the reasons is of course that the truth of the Spencer
business has become very apparent, so there is not a lot to
argue about. [Cookson, 32]

In quotation 6Qa, Cookson begins by defining the current debate over
stoichiometries, that is, over the number of protons crossing the
membrane, as crazy. It is crazy, he suggests, to engage in heated debate
because the implications for chemiosmosis are fairly trivial. Yet he goes on
to state almost immediately that the outcome of what he has called the
'battle over stoichiometries' will actually have major consequences for
participants' conception of the basic mechanism of energy transduction.
Thus, this controversial aspect of chemiosmotic theory is both important
and unimportant (6Q1–3). These remarks are confusing, if they are read
literally as referring to a single theory, namely, *the* chemiosmotic theory.
They become more understandable, however, in the light of the speaker's
subsequent adoption of two versions of chemiosmosis, one of which
excludes all those elements which he defines as still being scientifically
problematic (6Q9). Like the previous speaker, Cookson changes the scope
and content of the theory as he speaks in such a way that the essential validity
and general acceptance of one of these versions can never be in doubt.

Spencer's position on stoichiometries is often linked in the research
literature to his conception of protons being moved across the membrane
by a configuration of loops. Consequently, Cookson comments in passage
a on Spencer's formulation of the loop mechanism. He states that there is
very little evidence in support of that mechanism (6Q4). Nevertheless, he
claims, this does not mean that 'chemiosmosis' is reduced to the kind of
crude, basic version that we found in quotation 6P (6Q6–7). Chemiosmosis, he maintains, is more than the assertion that 'the proton gradient
has something to do with ox phos'. Cookson proposes that there is
something called 'the overall theory' which has actually been shown to be
correct (6Q8). He then suggests that this theory has a very detailed and
specific biochemical content (6Q9). However, he then formulates this
overall theory in such a way that the loop mechanism and anything to do
with the generation of the electrochemical gradient are expressly excluded
(6Q9). Furthermore, in sentences 10–11, he identifies the aspects of the
chemiosmotic theory with which he is concerned as those dealing
primarily with transport of substrates rather than with the manufacture of
ATP. Thus, for this speaker in these passages, chemiosmosis is presented
as some kind of general, yet detailed, theory which to a considerable
degree has been experimentally confirmed; yet, at the same time,

supposedly the same theory is interpreted in idiosyncratic terms as covering only those phenomena with which the speaker himself is centrally concerned and as composed only of those theoretical claims which he takes as validated.

Like Grant (6P), Cookson presents himself as a Spencerian or as in agreement with chemiosmotic theory, yet as differing in various significant respects from Spencer and as endorsing only a fairly narrow selection from Spencer's published claims. In quotations 6Qc and d, Cookson makes strong assertions about the degree of consensus with respect to chemiosmotic theory; and he goes on to use this supposed consensus to account for the lack of public argument over chemiosmosis in recent years. It is quite unclear, however, what Cookson means in these passages by such phrases as 'what Spencer has to say' and 'the Spencer business'. Which aspects of chemiosmotic theory is he referring to?

We suggest that in quotations 6P and 6Q, the speakers clearly alter the meaning of such terms as 'chemiosmotic theory' and 'Spencer's hypothesis' as they proceed. By varying the meaning of these terms, they are able to allow for the existence of a range of scientific disagreements among their colleagues, often with respect to apparently fundamental issues, and for marked differences between their own formulations and those proposed by Spencer in the literature, without giving up their claim that there is a substantial, even overwhelming, cognitive consensus. Grant achieves the appearance of consensus by reducing chemiosmosis to a basic, consensual version and by 'showing' that his own detailed differences with Spencer are not inconsistent with acceptance of this essential chemiosmotic concept. Cookson explicitly rejects the use of any grossly simplified version of chemiosmosis; a device which he has presumably encountered in the course of informal discussion with his colleagues. Instead, he is able to construct his strong consensus claim by blurring the distinction between Spencer's 'overall theory' and his own personal and partial interpretation of that theory. Cookson's use of the notion of 'the overall theory' is in practice equivalent to Grant's notion of 'the basic version'. Both concepts are employed to exclude from chemiosmotic theory whatever the speaker does not accept and whatever the other members of the field are said not to accept; whilst at the same time, both concepts are used as if they refer unequivocally to some theoretical entity, namely, the chemiosmotic theory, which exists independently of speakers' highly variable interpretations. Thus we can observe these speakers sustaining an appearance of consensus in their discourse through the subtle deployment of various versions of a theory which is said to be generally accepted.

Let us offer just one more illustration of the variable meaning of 'chemiosmosis'.

6R

(a) 1 I think people in general accept chemiosmosis up to a point: that electron transport certainly generates a potential across the membrane, certainly generates a movement of protons when measured under certain circumstances. 2 And the evidence certainly shows that a membrane potential or a proton gradient or both can make ATP. 3 But it doesn't hang together when you take it much further than that. [Milner, 22]

(b) 4 Spencer postulates a movement of the protons between the two bulk phases. 5 Now recently, a number of experiments have come out that suggest that the pathway for the protons is not between two bulk phases, and maybe only between one bulk phase and the membrane or maybe even within the membrane itself ... 6 So a bulk pH gradient across the membrane does not appear to be a high-energy intermediate state that's required to make ATP. 7 That was predictable anyway because Jarvis had earlier found ... [Milner, 23]

(c) 8 Spencer pointed out that you don't get oxidative phosphory-lation unless you have a complete vesicle. 9 Now that's been taken as an article of faith for many years and it made a lot of sense. 10 The only trouble is that the force of the argument now looks in retrospect not very great. 11 Because ... there is no way to make a piece of membrane that is not a vesicle. 12 The only way you can get any pieces of the whole system that is not a vesicle is to put it into a strong detergent. 13 The detergent replaces the membrane. 14 So ... you no longer have a membrane and you no longer have oxidative phosphorylation. 15 But that isn't a very good test, because you haven't got a piece of membrane any more and furthermore the detergent is inhibiting all these enzymes [which make ATP] ... 16 So the force of that argument [in favour of chemiosmosis] is now lessened a great deal. [Milner, 24]

(d) 17 Well, I think [these results] mean that the original form of the chemiosmotic hypothesis, that a complete membrane vesicle is required and that a proton gradient is required across the membrane, if these results can all be confirmed, they imply that what might be involved is a proton cycling. 18 *I would still regard this as chemiosmotic, although it's an unfortunate name then* [emphasis added]. 19 I would tend to want to rename the idea as an electro-chemical proton mechanism . . . 20 That's the way the wind is blowing currently. [Milner, 25]

These four fragments are taken from a longer passage in which the speaker identified a series of difficulties with the chemiosmotic hypothesis. Two of these are partly reproduced in sections 6Rb and c; and their theoretical implications are summarised in 6Rd. In 6Ra, the respondent offers a typical consensual version of chemiosmosis. He briefly summarises those aspects of chemiosmosis which people in general accept and his version, although short, seems roughly in line with Spencer's own published formulations; that is, it refers to electron transport creating a trans-membrane gradient which produces ATP.

In the two following sections, however, Milner brings into question two of the fundamental claims of chemiosmosis, namely, that a closed membrane (6R8–16) and a bulk gradient across the membrane (6R4–7) are required for ATP synthesis. He asserts that there are now good grounds for abandoning these features of chemiosmotic theory. The speaker recognises that further confirmation of recent experiments is required (6R17). But he suggests, unlike Spencer in the Nobel Lecture of the previous year, that the trend of opinion is currently moving away from these chemiosmotic assumptions (6R20). Nevertheless, he avers, he would still regard the processes involved as chemiosmotic (6R18). In making this claim Milner seems to stretch the meaning of the term chemiosmosis to its limits. He stresses that he is moving away from the central assumptions of chemiosmosis to such an extent that the very word no longer seems appropriate (6R19). Yet, he maintains, his approach is still in some sense chemiosmotic. Like the two previous speakers, Milner maintains an appearance of general acceptance of chemiosmosis by subsuming radically different scientific claims about specific biochemical processes under a highly general interpretation of that term.

Slightly later in the interview, having proposed further necessary alterations to chemiosmotic theory, he offered another consensual version of chemiosmosis. This version is even more basic than that offered in 6Ra. It is formulated in such a way that the essence of chemiosmosis can be seen clearly to include even the radical innovations which he has just recommended.

6S

Everyone accepts that the fundamental particles of oxidative phosphorylation are the electron and the proton. If there was nothing else to the chemiosmotic hypothesis than this it would still be a very important contribution . . . I think the whole fraternity working in the field feels that Spencer has been very doctrinaire in his attitudes towards his own hypothesis. When you look back on the history of many scientific hypotheses, they've all had to be modified in one way or another . . . [Milner, 26–7]

In this passage, the basic contribution of chemiosmosis is taken to be that of adding the proton to the electron as one of the fundamental particles involved in the production of ATP. From this point of view, any analysis of oxidative phosphorylation which includes both these particles can presumably claim to be chemiosmotic. There can be no doubt that this definition of 'the chemiosmosis which everyone accepts' is a far cry from the versions given by Spencer in his papers and lectures and in his interview. However, it is Spencer who is taken to task here for refusing to redefine his version of the chemiosmotic theory so as to bring it into line with the 'whole fraternity' of scientists working in the field. Spencer is criticised for being 'doctrinaire in his attitudes towards *his own hypothesis*'; that is, in this passage, for treating chemiosmosis as a concrete, detailed theory, rather than as a basic claim that protons as well as electrons are important. In this quotation, the speaker clearly treats 'Spencer's own hypothesis' as identical to the particular interpretation of chemiosmosis which he happens to formulate at this juncture. This respondent, like those quoted above, presents himself as uniquely able to speak on behalf of the theory in question. It is through *his* voice that the chemiosmotic theory which is coming to be agreed makes itself known.

In this section, then, we have seen how the meaning of such terms as 'chemiosmotic theory', 'Spencer's hypothesis', and so on, vary from one speaker to another. We have also seen that each respondent employs more than one interpretation of chemiosmotic theory in the course of the informal talk occurring in interviews. We have suggested, as a first step in analysing this kind of data, that we treat each actor as moving between an idiosyncratic version of chemiosmosis and a consensual version.

The interpretative variability of 'chemiosmosis' is not easily discerned in the ordinary course of events. Much of the time, it is hidden by the character of scientists' discourse about consensus. For researchers regularly speak as if 'chemiosmosis' is an entity held in common with most other colleagues. They each proceed as if the specific version of the theory that they are engaged in proposing is 'the real chemiosmosis' which is coming to be accepted or rejected by the field. They continually construct their accounts as if they are referring to 'a theory' which exists independently of their interpretative work. However, the detailed comparisons between accounts carried out above reveal that the apparent facticity of chemiosmosis and its apparently widespread endorsement are illusory in the specific sense that they exist, not as objective entities in an external social world, but only as attributes of participants' contingent consensus accounts.

Consensus as an occasioned interpretative product

The consensus accounts we have examined are always closely linked into the rest of the speakers' discourse.[6] For instance, Spencer used his consensus diagram in the early 1970s as a basis for revealing the dogmatism of some of his opponents as well as for prophesying about the future development of the field. His subsequent claim in the Nobel lecture that most of his major opponents had at last been converted to chemiosmosis, enabled him to characterise his erstwhile antagonists as basically altruistic and as willing, in the long run, to restrain the promptings of self-interest for the benefit of science. Others used assertions of a chemiosmotic consensus to explain why the field was closing down. Still others moved from denying that consensus to endorsing the scientific superiority of alternative theories and to identifying a range of emergent scientific problems which guaranteed that the field would be intellectually lively for years to come. Thus assertions and denials of cognitive consensus are important building blocks in scientists' discourse. They play a significant part in helping scientists to construct forceful and coherent characterisations of their social and intellectual world.

In the analysis above, however, we have been less concerned with studying how consensus accounts contribute to the meaning achieved in extended sequences of scientists' discourse, than with the interpretative structure of consensus accounts themselves. We have shown that Spencer's claim as formulated in his diagram could be read as contingent, as well as a literal description of the self-evident. Those scientists who challenged Spencer's claim drew attention to his 'inaccuracies' in identifying the membership of the field, in specifying individuals' scientific views and in describing the scientific content of the supposed consensus. These challenges made visible the three basic interpretative issues which can be seen to have been resolved in any consensus account.

We suggest that it is impossible for participants to furnish definitive solutions to these three interpretative issues. For such solutions involve unformalisable, practical judgements; judgements which are indirect, inferential and dependent on the particular interpretative context in which the judgement is being made. For instance, as we showed above, there is no single, unambiguous way of defining 'the field' in which our scientists work. Similarly, scientists' beliefs cannot be directly observed by their colleagues. Rather they are inferred from the published literature and from informal discussions, and then subsumed within a narrow range of conventional theoretical categories, the meaning of which appears to differ from one researcher to the next and from one occasion to the next.

This does not mean that scientists construct consensus accounts in a random fashion. Certain recurrent interpretative methods seem to be regularly employed in the accounts asserting consensus that we have examined. For instance, the following procedures appear to be in evidence in our material:

(a) treat each scientist as committed, at any given time, to a single scientific viewpoint or belief;

(b) treat each viewpoint as clearly evident in a scientist's written products and informal statements, yet as something separate from these products;

(c) treat each theoretical label as having a clear, invariant meaning;

(d) treat the view of (most) individual scientists as coinciding with one of the current theoretical labels;

(e) employ consensual and idiosyncratic versions of a theory so as to reconcile cognitive variation with the existence of consensus.

It is tempting to refer to these recurrent features of scientists' consensus accounts as resulting from the existence of a widely shared scientists' 'folk theory of cognitive consensus'. It seems to us, however, that the notion of 'folk theory' should be avoided in this case.[7] For the features we have identified are not explicitly stated by participants themselves. They are, rather, analysts' formulations describing certain interpretative procedures[8] which seem to occur regularly in a collection of accounts. There is no evidence to suggest that they depend on a theory held by participants. Thus, the features summarised above are best seen as recurrent interpretative procedures which are embodied in the collection of accounts under investigation, in the sense that they can be made visible by the kind of systematic comparison we have adopted.

These general procedures (a–e) appear to help scientists resolve the three underlying interpretative issues. Indeed, they can be described as ways of reducing or concealing interpretative contingency. For example, in treating the attribution of individual belief as unproblematic in giving an account of consensus, the scientist is ignoring those occasions, which are endemic in informal interaction in science, when scientists experience enormous difficulty in comprehending one another's technical arguments. Similarly, by treating each theoretical label as having a clear, invariant meaning, participants create an aura of facticity for each theoretical position and convey the misleading impression that those scientists subsumed under a given label actually endorse the same set of scientific beliefs. Nevertheless, although these procedures enable scientists to disregard huge areas of interpretative contingency, they do not place any narrow restrictions upon the precise content of consensus accounts. For, as we have seen, not only do the consensus accounts of different scientists differ considerably, but the accounts produced by a given scientist vary from one occasion to another. In this sense, consensus accounts can be seen to be 'occasioned', even though they reveal certain recurrent interpretative features which can be observed and formulated by the analyst.

It would be misleading, therefore, to interpret consensus accounts as following inevitably from participants' resolution of the three basic issues by the application of a set of determinate rules to particular cases. This was seen most clearly above in connection with Spencer's consensus diagram. We showed in that case that there seemed to be no readily identifiable procedures for identifying membership, attributing belief, and so on, which *led* Spencer unavoidably to end up with a smooth upward curve of consensus. Thus, scientists' consensus accounts are neither literal descriptions of an independent social reality, nor are they the necessary outcome of scientists' standardised interpretative procedures. They are, rather, the means by which scientists make available to us, and to their colleagues, versions of the state of collective belief which are appropriate for specific interpretative occasions.

Analytical implications

Scientific consensus has been treated by sociologists as a typical collective phenomenon, that is, as a potentially measurable aggregate attribute of social groupings which exists separately from the interpretative activities of individual participants. Nevertheless, empirical study of scientific consensus clearly does depend on individual scientists' interpretative products. In the extreme case, like that of Knorr's study mentioned above, the sociologist establishes the degree of consensus simply by aggregating the consensus accounts of a number of individual scientists. But even less direct studies of consensus, for example, those using review articles or citations, depend unequivocally for their conclusions on scientists' own, context-linked and potentially variable symbolic products.[9]

In this chapter, we have examined several kinds of symbolic product which could plausibly have been used by sociologists as indicators of scientific consensus in 'ox phos'; for example, a review article, two honorific lectures and sections from interview transcripts. Moreover, the Nobel Prize has recently been awarded, a particular theory is coming to dominate the textbooks, prior theories appear to have been widely repudiated and strong consensus claims can be found in the interviews and elsewhere; all apparently clear signs of the existence of consensus in this research area. Yet we have shown that participants' consensus accounts are highly variable and that their meaning is linked to the interpretative situation in which they occur. We have shown that the consensual character of 'ox phos' can be either constructed or deconstructed, not only by different participants, but also by the same participants as they engage in new interpretative work. In view of these observations, it appears that, for the purposes of sociological analysis, a given field at a particular point in time *cannot be said to exhibit a specifiable degree of consensus*. Rather, the field must be said to exhibit varying degrees of consensus, depending on the discourse of those involved.

Scientific consensus, then, is neither distinct from members' discourse nor is it open, even in principle, to definitive measurement at any specific juncture in a field's history. Consequently, traditional analysis of this topic is doomed to failure. If consensus is open to various construals, there is no point in trying to show how a range of other social factors vary in accordance with *the degree* of consensus. Unlike traditional analyses, however, the form of interpretation we have begun to develop above is not undermined by the variability of discourse about consensus. For we focus analytically, not upon the highly variable consensus claims produced by participants, but upon the recurrent interpretative methods whereby variable symbolic products, such as consensus accounts, are contextually generated.

It is important to recognise that the interpretative procedures which we have identified are not the personal interpretative achievements of individual scientists; even though each text or utterance in which these procedures appear *is* a unique product. The objective of our analysis has been to identify recurrent, regularly used, and in this sense collective, cultural resources which are embodied in and visible in participants' discourse. Thus, in our treatment of the construction and deconstruction of consensus in 'ox phos', we have not been trying to replace the traditional analysis of collective phenomena with an individualistic perspective. Rather, we have been developing an alternative and more fruitful approach to the investigation of *social* regularities.

7

••

Working conceptual hallucinations

The analysis presented in previous chapters has two clear limitations. In the first place, it has been concerned primarily with discourse occurring within the research community. Secondly, we have so far concentrated exclusively on verbal discourse. An omission of non-verbal material would be a serious gap in a form of analysis dealing with scientific discourse, rather than with action and belief, because there can be no doubt that technical communication in science relies heavily on pictorial and mathematical repertoires.

Our objective in this chapter is to begin to overcome these limitations. We will do this by examining a set of pictorial forms produced by bioenergeticists, by studying how these pictorial forms change as they are made available to non-specialists, by investigating a particular non-specialist's reading of one of these pictures, and by looking at what our respondents have to say about this collection of pictorial products and about pictorial representation in general.

Scientists use many kinds of visual display, such as tables, graphs, photographs, electron micrographs, drawings, flow diagrams, and demonstrations, in communicating their knowledge-claims.[1] Here, we will focus on those displays which represent the chemiosmotic processes of ATP synthesis and the associated biological structures. We will refer to such displays simply as 'pictures'. It is clear that these pictures are closely related to verbal formulations of the chemiosmotic theory and that they play an important part in the communication of that theory among bioenergeticists and from bioenergeticists to non-specialists. Chemiosmotic pictures, then, in some sense embody the chemiosmotic theory and reproduce it in visual form.

The analysis of pictorial discourse which follows links up in various ways with what has gone before and with the content of the next chapter. Because we are going to examine numerous pictures which appear in written texts, we will return to the type of analysis which was predominant in chapter three. In that chapter, we showed how verbal discourse differed systematically between formal and informal contexts. In this chapter, we distinguish between the context of the research literature and that of textbooks and popular presentations of scientific knowledge. We will show that there are systematic changes in the forms of pictorial discourse employed in these contexts.

In later sections of this chapter, we will examine numerous quotations from our interviews. In twenty-five of these interviews scientific pictures were explicitly discussed, often at some length. Our procedure was to ask each respondent about his own use of pictures and then to present specific pictures for more detailed comment. The most frequently discussed of these pictures appear below (pictures v to viii). Unlike many other topics covered in the interviews, reference to pictures occurred as a distinct and separate topic at the direct instigation of the interviewers. It is necessary to mention this, because it may be at least partly responsible for the considerably greater degree of consistency and coherence exhibited in respondents' talk about pictures compared with that found in their talk about most other topics. We must emphasise that, although our respondents only began to reflect on the nature of pictorial discourse in response to our specific requests, many of them *used* pictures of various kinds with no encouragement from us in order to convey their views on the biochemical processes of oxidative phosphorylation. As most sociologists who have interviewed scientists about their work will probably testify, interviews in which the respondent employs no pictures at all can be regarded as rather unusual.

Some typical pictures from bioenergetics

In order to appreciate some of the distinctive features of pictures appearing in bioenergetic texts, let us begin by considering one which is in some respects unusual. Picture i below is the frontispiece to a textbook on biological membranes and their cellular functions. This book is described in its preface as presenting 'a broad view of the significance of membranes in cellular activities, particularly for use by students and teachers in biochemistry and other biomedical sciences'.[2] The dual representation in picture i of the typical cell in animals and plants provides an overall context for the contents of the book. In subsequent chapters the authors deal systematically with the components it shows. The picture furnishes its readers with a preliminary idea of what each of the components of a cell 'looks like', how they differ from each other and how they are distributed within the cell. It provides an initial visual *Gestalt* which is filled out in detail as the chapters unfold.

Even though the scope of picture i is restricted to the cells of higher organisms, it is very unusual in representing so many different biological phenomena together. It illustrates a type of comprehensive picture which

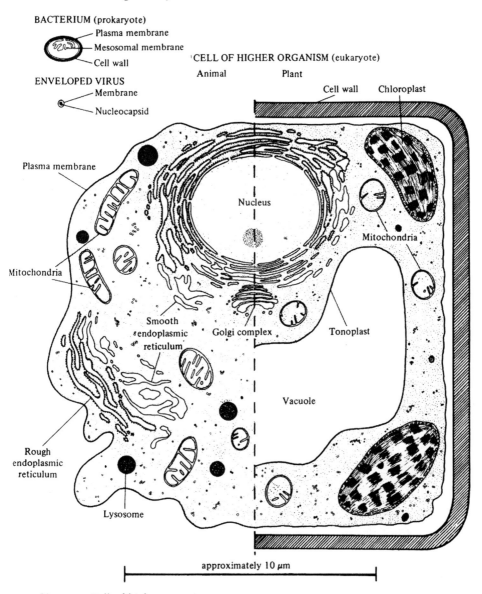

BACTERIUM (prokaryote)
Plasma membrane
Mesosomal membrane
Cell wall

ENVELOPED VIRUS
Membrane
Nucleocapsid

CELL OF HIGHER ORGANISM (eukaryote)
Animal   Plant

Cell wall   Chloroplast

Plasma membrane

Nucleus

Mitochondria

Mitochondria

Smooth endoplasmic reticulum

Golgi complex

Tonoplast

Rough endoplasmic reticulum

Vacuole

Lysosome

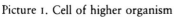

approximately 10 µm

Picture 1. Cell of higher organism

hardly ever occurs outside textbooks. The scientists we have studied appear to treat the wider biological context portrayed here as entirely irrelevant when constructing their own pictures. Our scientists are concerned with processes occurring in the inner membranes of the small particles referred to as 'mitochondria' and 'chloroplasts' in the diagram above. (Some of them are also concerned with bacteria, pictures of which we shall discuss below.) Their pictures, therefore, typically represent one minute segment of such membranes. In addition, they depict in a highly conventionalised form just a select few of the processes and components assumed to operate in these segments. Even when these scientists are communicating with students or laymen their pictures usually retain this very narrow and highly selective focus. This is exemplified in picture II, which depicts the mitochondrial membrane along with the basic processes that move protons across the membrane.

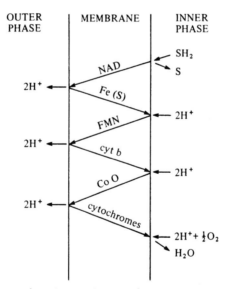

Picture II. Proton-translocating respiratory chain

This picture is from the research journal *Advances in Enzymology*, but almost identical pictures appear in biochemistry textbooks throughout the 1970s. One of its obvious characteristics is that, unlike picture I which resembles in certain respects what one might observe by means of an electron micrograph, picture II gives a minimal impression of realistic representation. For instance, not only is the cellular environment of the mitochondrion totally ignored, but also its outer membrane. For our respondents, such phenomena are consistently treated as irrelevant to the topic of oxidative phosphorylation. Accordingly, these phenomena almost never appear in their pictures. However, picture II is even more narrowly focused than this. Virtually no information is provided about the structure of the inner membrane. It is represented by two parallel straight lines, even though it is 'known' to be exceptionally convoluted. The symbols for various molecules (Fe(S), FMN, etc.) are placed within the boundaries of the membrane, but no indication is given about their precise topography. For instance, the picture does not show whether they are on the surface of the membrane, fully in the membrane, partially embedded in the membrane or spanning the membrane; or whether they are structurally contiguous or structurally separate, and so on.

Such a picture, despite its abstraction and selectivity, can be further simplified and condensed, as in pictures III and IV. In picture III, the content of picture II is contained in the top line and the first downward curving arrow. By means of extreme abstraction, simplification and selectivity, picture III is able to summarise the whole process of ATP production. Picture IV has the same scope, but it provides greater detail than III and it introduces a clear indication that ATP synthesis arises out of the movement of protons back and forth across an organised membrane. In picture IV, the content of II is condensed into the top three curving lines.

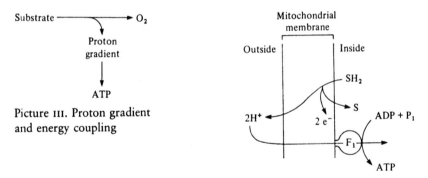

Picture III. Proton gradient and energy coupling

Picture IV. Chemiosmotic coupling

Although we have only looked at and commented briefly on four pictures so far, let us summarise some of the formal characteristics of such pictures. These characteristics will be further illustrated by pictures appearing in subsequent sections.

Some formal characteristics of scientists' pictures

We suggest that the great majority of pictures used by research scientists and by textbook authors in relation to bioenergetics and the chemiosmotic theory of oxidative phosphorylation have the following general features:

(1) *Generality*. Pictures refer to generalised phenomena rather than to specific, observable entities. For example, we find countless pictures of *the* mitochondrial membrane, but no attempt to depict any particular cell's membrane. This may not be equally true of all scientific disciplines. Rudwick has shown that reproductions of particular landscapes have played a part in the 'visual language of geology'.[3] Such unique representations seem to be completely absent from the literature on bioenergetics. We suggest that it is likely to be geology rather than biochemistry which is unusual in this respect.

(2) *Selectivity*. The great majority of pictures deal with very specific research issues. Even when pictures summarise a considerable body of research, they re-present this material in relation to some analytically defined issue. Pictures are very seldom designed to depict phenomena in their full 'naturally occurring complexity'.

(3) *Conventional simplification*. A limited range of standardised forms is employed; in particular, straight and curved lines, arrows, circles and boxes. These conventional resources are used, not only to depict actual biochemical processes (for example, the movement of protons across membranes), but also to simplify and make unproblematic aspects of the phenomena which are not in question at a particular juncture. For example, by representing the surface of the membrane as a continuous straight line easily transversed by 'proton-carrying' arrows, an illustrator is eliminating from consideration any question of relevant interaction between its surface and the phase outside.

(4) *Conceptual reference*. The pictures do not refer directly to empirical phenomena but to conceptual entities or idealised versions of observable phenomena. This follows from the three prior points. For example, it is impossible actually to observe the generalised membrane represented in pictures II and IV or the joint animal–plant cell in picture I. These pictures are composite constructions based on various observations of particular cells and membranes together with inferences from a range of experiments involving numerous particular biological objects. This point corresponds to Ravetz's portrayal of the conceptual language of science as dealing with 'intellectually constructed classes of things and events'.[4]

(5) *Interpretational variability*. Pictures are part of and are embedded in a conceptual argument. Accordingly, the nature of the picture changes as the argument changes. Thus III and IV are adjoining figures in the same text. Picture IV is an elaboration on picture III which takes for granted the central point expressed in III and specifies that point in

further detail. However, the relatively comprehensive picture IV contains much less specific information about proton translocation than does picture II, which deals solely with the latter topic. The character of the pictorial representation, then, varies in accordance with the interpretational work being carried out in the text as a whole.

(6) *Contextual variability.* Because scientists' interpretative work tends to vary from one social context to another, pictures are also to some extent context-dependent. For example, pictures in research reports tend to differ from those in reviews. And pictures in textbooks and popular accounts of recent developments, although in some cases copied directly from the research literature, are often supplemented by pictures devised specifically for the context of teaching. This is true of pictures I and III above and of VII, VIII and IX below. Thus, in so far as pictures can be said to embody scientific knowledge-claims, detailed variations in pictorial form and content serve to reveal the context-dependence of such claims. This is not to suggest that the different pictures produced for different contexts are necessarily inconsistent or incompatible. It is to suggest rather that participants produce different versions of their knowledge for different contexts and that scientists' capacity to extract from these versions an ultimate formulation of 'what the pictures really mean' depends on complex interpretative skills which are not always shared with outsiders.

(7) *Interdependence of visual and verbal texts.* This interdependence is most clearly evident in the regular inclusion of textual symbols within the picture. This can be seen in all the pictures above, which include such symbols as 'mitochondrion', 'membrane', 'proton gradient' and 'ATP'. In addition, there is a more subtle kind of interdependence in that, although pictures are spatially separated from the written text and usually clearly labelled as distinct entities, they are typically presented as a summary or illustration of what the words mean. Pictures tend to have an unwritten, implicit heading along the following lines: 'In other words, what I have been saying [or what I am about to say] looks *basically* like this.'

(8) *Non-reflexivity.* Pictorial representation in science is overwhelmingly a form of non-reflexive discourse. For instance, scientists' representations of biological membranes contain various circles, lines, blobs, arrows, etc., which refer to generalised versions of phenomena in living organisms which are supposedly observable in a variety of direct and indirect ways. However, the meaning of these shapes and forms, for example, what connection they are to be taken as having with real membranes, cannot be specified in terms of the shapes themselves. Unlike verbal languages, the resources of which are used routinely to

consider the meaning of particular verbal statements or even language itself, visual languages of the kind used in science appear to be primarily unidirectional. They seem to point rigidly beyond themselves towards the objects and processes in the natural world which they represent. Thus the nature of the visual language of science can only be portrayed in verbal terms. Yet very few verbal instructions are provided in scientific texts to guide readers' interpretation of pictures. There are occasional references in textbooks to the 'schematic' character of specific representations. But such remarks convey little positive guidance. It is not made clear, for example, whether other pictures which are not so labelled are to be taken as 'non-schematic'; whether the entire picture is equally schematic or whether certain components are more realistic than others, and so on. On the whole, the interpretative practices for 'reading' pictures are left to the readers' discretion and to any clues which he can extract from the verbal or visual text. (Towards the end of this chapter we shall examine in detail an unusual picture which *does* appear to give directions for its reading.)

Pictures and vocabularies of verbal discourse

The eight characteristics of scientific pictures given above are neither a comprehensive list, nor are they necessarily applicable to every pictorial representation produced by scientists. They are rather some of the more obvious features of the great majority of pictures used in bioenergeticists' research papers and reviews and in the appropriate sections of biochemical textbooks. There is no reason to expect, however, that they are unique to this area of research.

In this section, we will try to draw out some sociological implications of these eight points by examining scientists' talk about four further pictures (v to viii). Most of the discussion of specific pictures in our interviews dealt with these representations. Picture v is taken from a review paper in *Biochemical Society Transactions* which focuses on the issue of how many protons cross the mitochondrial membrane to create each 'high-energy phosphate bond' (i.e. each unit of ATP). Picture vi is also from a review paper, but one which deals with the broader topic of the structure of biological membranes. Thus this picture is a generalised portrayal of the components and organisation of an unspecified membrane. Picture vii is taken from an article entitled 'How cells make ATP' which appeared in *Scientific American* in 1978 and which furnished a strongly chemiosmotic review of the processes of oxidative and photosynthetic phosphorylation. Like pictures ii, iii, iv and v, it represents the membrane of the mitochondrion. Picture VIII is from the same *Scientific American* review and deals with the production of ATP in the membrane of the bacterium *E. coli*. Thus, the last two of these pictures were published in a journal available to and regularly read by non-bioenergeticists. We will see that these pictures prompted our respondents to talk about the topic of communication with non-specialists and about non-specialists' possible misinterpretations of such pictures.

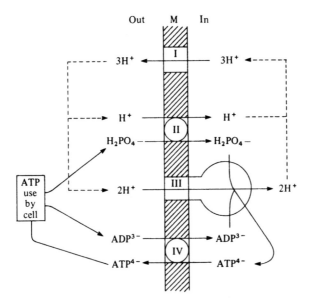

Picture V. Oxidative phosphorylation in mitochondria

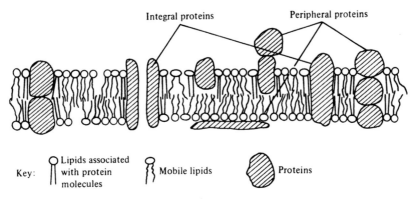

Picture VI. Model of a biological membrane

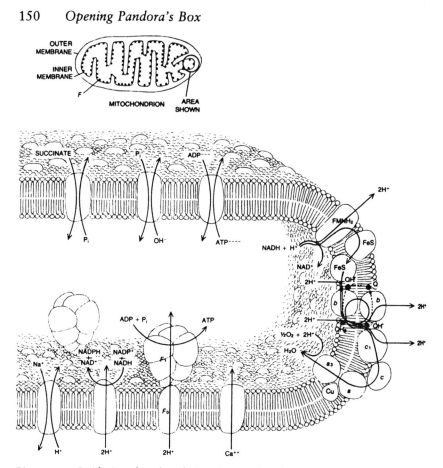

Picture VII. Oxidative phosphorylation in mitochondria

Let us begin our analysis by referring back to the property of non-reflexivity identified above. It follows from this property that participants' attempts to make sense of their pictures, to give an account of their meaning, must be carried out principally in verbal terms. We would expect, therefore, that participants' interpretations of their pictures would draw on the two basic verbal repertoires identified in previous chapters. Furthermore, whereas the empiricist vocabulary portrays the cognitive products of science as literally representing invariant features of the natural world, scientific pictures typically represent the world by means of a discourse which is selective, conventional, interpretative and variable. We would therefore expect the characteristics of pictorial discourse to be more closely aligned to the contingent repertoire. The contingent character of scientific pictures is in fact strongly emphasised in our

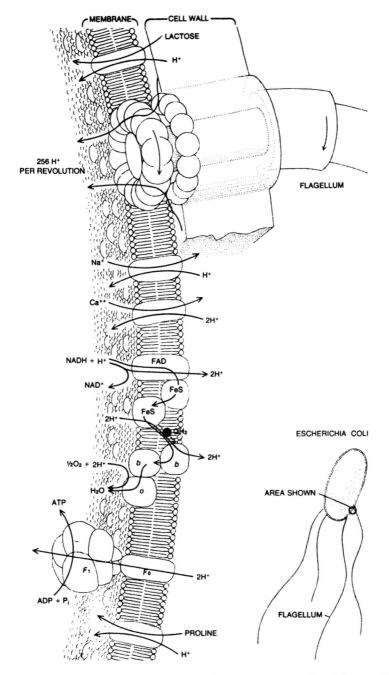

Picture VIII. Oxidative phosphorylation in the bacterium *E. coli* and the rotation of the flagellum

respondents' informal talk about pictorial representation. Our respondents' characterisations of pictures v to viii and of other equivalent pictures can be placed along a continuum varying from 'complete fictions', which are made to appear as totally contingent cultural products, to 'completely realistic representations', accounts of which are couched in strongly empiricist terms. The great majority of characterisations in our transcripts are located towards the 'fictional' end of this continuum.

In the following discussion, we will continue to use the terms 'realistic' and 'fictional' as these terms seem more appropriate than empiricist or contingent in view of the fact that participants are talking about pictures and not about action or belief. Nevertheless, it must be stressed that there is a marked parallel between scientists' realistic and fictional talk about pictures and their empiricist and contingent talk about action and belief. We will proceed by presenting and commenting on a representative sample of interviewees' characterisations, beginning at the more 'realistic' end of the distribution.

Only one of the 25 respondents with whom we discussed pictorial representations made unequivocally realist claims for any particular picture. In the following extract, the respondent concludes a series of critical remarks about picture vii by contrasting the speculative content of that picture with his own accurate representations.

> **7A**
> This is a kind of [chemiosmotic] way of representing, you don't, you just dream up some way of doing it and you do it this way. But *this* [*pointing to his own picture*] is the reality . . . It's a reality, you can see it. And they just ignore it . . . I am not talking about what we *might* see or could see, but what we *do* see. [Pugh, 49–50]

The organisation of this passage resembles that found in accounting for error. The speaker's pictures are directly equated with the observable world, whilst picture vii is dismissed as something merely 'dreamed up'. Other speakers, however, did not accept this characterisation of Pugh's pictures. They described them, for example, as 'imaginative' and 'not based on fact'.

Although no other respondents made such a strong realist claim for any existing pictures, one scientist suggested that it may be possible to draw realistic pictures of the phenomena of oxidative phosphorylation in due course.

> **7B**
> *Fasham:* There is some speculation in here [picture vii] but much of it is built on experimental fact ... With any representation it is an approximation ... This picture of the membrane with all the bumps over

it, that's pictorial, that's not factual . . . It is trying to put things as realistic as possible, but it is still schematic in nature . . .
Interviewer: Do you think it will ever be possible to draw a picture which *is* realistic?
Fasham: Oh sure. It will be eventually I hope. Sure . . .
Fasham: You have got to always remember they aren't *the* truth. Whatever *the* truth is, that's almost certainly not it. [Fasham, 30, 36]

In this passage, the respondent offers a verbal estimation of the *degree* to which picture VII is realistic. He stresses that certain features are derived from experiment and observation, but that other elements are speculative or schematic. When the interviewer asks about the possibility in principle of realistic representation his answer, unlike that of other interviewees to this question, is strongly positive. Yet in the penultimate sentence of this quotation, which occurs several minutes later in the interview, he appears to assert emphatically that pictures are never entirely realistic: 'You have got to *always* remember they aren't *the* truth!' Similarly, he has stressed earlier that all representations are approximations. Thus this speaker's characterisation seems to be that pictures can be assessed fairly unambiguously for their degree of realism and that it is possible in particular instances to decide which of several pictures is the most realistic, even though no completely accurate version is at present available or even possible in principle. He also appears to claim that the degree of pictorial realism in this field is likely to increase over time, without necessarily ever arriving at a completely accurate representation of the phenomena in question.

Four of our biochemists, including the respondent quoted just above, talked explicitly in terms of pictures being more or less realistic, or representing phenomena more or less faithfully. But these assertions were in each case qualified by references to the speculative, hypothetical or fictional components in pictures. This was true, for example, of the two authors responsible for picture VII. In the following quotes, they are discussing that picture.

7C
There are conventions certainly for representing membranes now. You just draw the little phospholipids and then everybody knows that means that it's fluid and it's all of the various things of the fluid mosaic model [represented in general form in picture VI] in effect. And often proteins are written as circles and squares. But we felt that the one thing we know about them is that they are not circles and squares. So we tried to make it a little more realistic looking. But we simplified it considerably in the pictures that we made in *Scientific American* . . . It's just a question of putting in, the chemiosmotic theory really, into what it should look like

... These things are not really for posterity so much. They're the best guess that we have at the time or they're for purposes of simply making things clearer even if it's wrong. At the time I was writing [a particular feature in one of the pictures] I didn't really think it was right, but it was just a way to talk about it. [Smith, 54, 59 and 65]

7D

[We tried] to make a model that really had some basis in reality. You know, you can always draw these lines and put boxes on the side, but we even tried to keep the size of the proteins roughly to scale . . . [But a lot of] this kind of information we don't have yet. So it's 'more realistic', in quotes – a more detailed view of the way it could be. [Trubshaw, 91 and 95]

Both these authors present picture VII as being more realistic than other, more conventional, portrayals of the mitochondrial membrane. They are comparing it here with pictures such as V above. In their picture, for example, proteins are said to be given more realistic shapes and to be drawn approximately to scale. Yet at the same time, Trubshaw emphasises that much of the information necessary for an accurate picture was not available. Hence a great deal of the picture was a speculative view of what the membrane and its constituents *could* look like. The other author similarly refers to the picture as the best guess possible in the circumstances. He describes it as a way of exemplifying how the membrane *should* appear, given the validity of the chemiosmotic theory. And he admits to including one incorrect element in the picture simply in order to maintain a coherent overall presentation.

There is nothing necessarily inconsistent in these authors stressing both the inaccuracies of their picture and its greater degree of realism. They are presenting picture VII and picture VIII as an improvement in this respect on the standard, highly conventionalised, schemes current in the research literature and as realistic as can be expected in view of certain practical limitations. Thus these authors, like Fasham above, provide a realist account of their picture in the sense that its representational inadequacies are treated as a practical problem and as one which is, therefore, potentially resolvable over time. Yet at the same time, they recognise that the pictures they can actually draw at the moment contain a very significant fictional component.

In contrast with Trubshaw and Smith, five of our respondents explicitly stated that picture V was preferable to picture VII precisely because it *was* obviously conventional and made no pretence of realistic representation. These scientists tended also to stress the fictional character of pictures generally in this field. The passages quoted below come closer than those quoted above to treating the unreality of pictures as a feature which has to be explicitly recognised and taken into consideration when using them as a means of communication.

7E

Diagrams are dangerous, because drawing a diagram of *this* kind [*pointing to picture* VII], a pictorial representation rather than a reaction scheme, implied a physical reality and a permanence about the concept. I actually prefer [v-type] diagrams, because whether they are right or wrong they are formal, entirely formal representations of entirely hypothetical reaction pathways with a topology . . . [Picture v] is a better representation because it doesn't imply that it's telling you how things work. [Harding, 61–2]

7F

If you have been in this field for a while, you have seen a lot of pictures come and go and you take them all with a grain of salt. But I think figures, visual aids, are very important to help you grasp what the author is trying to get at . . . I think this [picture v] might be a little better because this is *obviously* a scheme and that [picture VII] pictures it as really what the membrane might be. Maybe it is a good figure. This one is honest in that it, I think, conveys to me immediately that this is a scheme and no attempt to state the actual. [Hargreaves, 72–3]

7G

[Picture v] is more abstract. This [picture VII] attempts to give a *pictorial* representation . . . It is drawn as if one were a molecular-sized entity taking a look at the field . . . I think that probably at the molecular level the interaction of light and matter is not sufficiently clear to give you pictures that look like that. I think it's not appropriate. [Miller, 52–2]

These researchers, then, propose that pictures, in this field at least, cannot be regarded as realistic representations; that they should therefore be clearly and explicitly conventional in form; and that the more apparently realistic a picture is the more misleading it is likely to be. In the first two passages (7E, 7F), the speakers do not state whether this is merely a temporary state of affairs, which could be due to the intellectual immaturity of the field and which might change as the corpus of established knowledge grew, or whether it is a permanent characteristic of pictorial representation. The third speaker (7G), however, seems to treat the conventional nature of pictures in this field as unavoidable in principle. Few of our interviewees addressed this issue directly. It is not possible, therefore, to assess how far our respondents in general were treating the fictional nature of pictures as intrinsic to the phenomena of bioenergetics or intrinsic to the realm of pictorial discourse. What is clear, however, is that the great majority of them, that is, twenty out of twenty-five, emphatically characterised pictorial representation in their field up to the date of the interview as unavoidably speculative, hypothetical, uncertain, interpretative, highly personal, and so on.

7H

(a) I think, on the whole, people know them [pictures] for what they are and probably don't take them too seriously. [Holloway, 54]

(b) We're working with the light harvesting systems which aren't even *in* this model . . . So you can't take a model like this seriously. [Grant, 68]

(c) [Pictures v to viii] are *all* so far away from the truth that the difference between them is really rather small, compared with the difference from the truth. [Norton, 59]

(d) I think they're such personal things. They represent in fact a way that you yourself are thinking about what's going on. And I often think they're highly individual to that person or to a group of people who happen to use the same type of symbol. [Burridge, 31]

(e) God knows what [these molecules] look like . . . [Pictures] are bound to be inaccurate and there's no way of telling how inaccurate. It's a mental picture put down on paper, but put down in a comprehensible – an attempt to make clear what might be happening. That's the best you can say. [Richardson, 22–3]

(f) We have two slides of what we call 'lies and schemes', most of which are redundant. But we keep them to show how we are changing . . . you haven't a clue what it really looks like. [Jeffery, 63 and 67]

(g) I think [picture vii] does give you some sense of reality. But I think you've got to be careful of assuming that *is* reality. As long as you know that it's just a view of what we have now, it's good. It could be nothing to do with the reality of the situation. [Scott, 71]

(h) Generally in these biological fields there is a great deal of pictorial representation which is often very misleading. Look at *Scientific American* for example. There are all these biological papers with nice keys and locks and beautiful shapes drawn. And I often wonder, do the people who read these papers really believe that those things look the way they are drawn there or do they realise this is just supposed to be a pictorial representation of the truth which brings out those features of the problem that the author is trying to explain, and other than that it is totally fiction? . . . [*Picture vii is presented at this point*]. This picture is along those lines . . . Nobody knows that these things really move this way. This is science fiction . . . There is no 'reality'. There are so many 'realities' all depending on what you do about it, that there is no unique answer to the question. There are various techniques for getting pictures, freeze etching and the other techniques, and they would all give you different kinds of [pictures]. [Hinton, 15–19]

(i) I say that pictures are 'working conceptual hallucinations'. Nothing limits you when you make a picture. [Cookson, 80]

In these representative quotations, pictures are said to be 'far away from the truth', 'not to be taken seriously', 'very individual things', 'inaccurate to an unknown degree', 'lies and schemes', 'perhaps nothing to do with reality', 'science fiction', and even 'hallucinations'. Participants emphasise the fictional character of pictures v to viii, of other pictures in their field and occasionally of pictures in biology at large. The relationship between the biological phenomena under study and their pictorial representation is treated as highly contingent. As we will see in the next section, however, this relationship is not regarded as entirely arbitrary. For pictures are regularly described as being devised in particular ways in accordance with the requirement of specific social contexts. In particular, our respondents stressed that different *degrees of realism* are appropriate, depending on the audience for which the text is intended.

The context-dependence of pictures and Trubshaw's dilemma

Our respondents' generally fictionalist treatment of pictorial representation is clearly evident in their remarks about the production of pictures for students and for other non-specialist audiences. As we noted in a previous section, one of the formal characteristics of scientific pictures is that they tend to vary from one context to another; in the sense that, although pictures from research papers do reappear in textbooks and popular presentations, they are accompanied by significantly different kinds of pictures which appear only in the latter texts. A clear verbal rationale for such contextual variation is provided by our sample of researchers. In the first place, it is maintained that pictures often have to be devised specifically for non-specialists because the latter are not properly equipped to understand the esoteric forms of communication employed within the research network. Secondly, it is said that such pictures have a distinctive aim; namely, that of conveying efficiently a *general impression* of the kinds of scientific processes at work. Thirdly, it is asserted that the detailed accuracy of such pictures is not important. Students and other non-specialists will not remember the details. The point of the picture is said to be that of providing them with a coherent overall presentation which will communicate the central scientific principles in operation. Finally, it is regularly proposed that it may be necessary to create an illusion of pictorial realism in order to communicate effectively with this kind of audience; or that an apparently realistic style of representation may somehow be most appropriate for this social context. This type of account of the role of pictures in communication between the specialist community and outsiders was clearly evident in eighteen out of twenty-five interviews. It is illustrated in the quotations which follow, in most of which the speakers are referring initially to picture vii.

7J

(a) This model here is acceptable. But if I wanted to be really critical, I'd say [*the speaker identifies a series of errors and shortcomings of picture* VII] . . . But it's not meant to be like that. All it's meant to do is to show that there is an electron transport chain organised in a certain way and that here is a phosphorylation site which is at another part of the membrane and you're utilising here hydrogen ions . . . It's really just a very superficial thing meant for the layman, really. And there's nothing wrong with that. [Grant, 71–2]

(b) [Picture VII] is too media-oriented. It's OK for somebody who's naive in the field and who wants to get the feeling that this is part of a real thing. And for the *Scientific American* it would be quite, would perhaps be better to do that [VII] than that [V]. Picture V to the *Scientific American* reader wouldn't mean anything. It's a formal reaction pathway. Picture VII is saying 'well, this really is what it looks like'. [Harding, 62]

(c) That's quite standard sort of *Scientific American* presentation. They've taken liberties, obviously, but they know that, they've said that . . . [*various inadequacies are specified*]. But I think that doesn't matter, for this level of article is after all a popularising one. They're not meant to be *that* critical. They give a clear idea of chemiosmosis from which a student or people like yourselves from outside can build. [Thompson, 3 and 5]

(d) You can explain the *bones* of [the chemiosmotic account of oxidative phosphorylation] to an undergraduate with a piece of paper and pencil now. So, in fact, the principle, the explanation of the principles has come down to a single diagram . . . This sort of diagram [picture VII] doesn't help *me* at all. And it could even be misleading to those who read the article. But if it enabled them to understand the *principles* of something better, then it would have served the purpose – you must be very careful. [Barton, 55 and 62]

(e) [Picture VII] is really an exaggeration of what we know exactly. If it were in a research paper I would not have done it quite like this. [Smith, 65]

(f) There are uncertainties there. There are things left out . . . But those are the general principles and for the purposes of that article it's fine. [Miller, 49–50]

(g) [Various aspects of picture VII] are absolute hog-wash. Those fatty acids are undoubtedly interlaced, so the picture is wrong in respect to that detail. So there are uses; it's alright, this must be from *Scientific American* I take it, and that's alright to introduce concepts like that to semi-lay readers I think. Articles in there are read by people outside the immediate field of interest usually . . . I think it's alright for lay readers. I think it's very bad for, if you were using it as a didactic presentation to students. It could thwart their efforts to find out exactly what is the case. [Waters, 38]

(h) I think it [picture VII] is marvellous for an audience of just students. I don't think it does much for third-year students or research workers . . . I would not embark on this sort of exposition unless I was talking to this audience. [Roberts, 42]

In these quotations, the variable nature of scientific audiences is emphasised and the need to construct pictures in a way which is appropriate to a particular audience is stressed. There are occasional differences of opinion about the precise audience for which picture VII is suitable (e.g. quotations 7Jg and h). But a generally fictional and socially variable account of pictorial representation in the area is maintained with marked uniformity.

Many of our respondents, having asserted the context-dependence of scientific pictures along these lines, went on to consider whether their non-specialist audiences would be likely to recognise the fictional character of these pictures. In many cases, our interviewees maintained that there was a danger of 'misinterpretation' when pictorial representations were devised for such audiences. In other words, having given an account of pictures in fictionalist terms, speakers often proceed to draw attention to the distinct possibility of others making sense of pictures in terms of the dominant realist or empiricist interpretative repertoire of science and thereby misunderstanding their true character. There are many examples of this, in addition to those already given above (7E, 7Hg and h, 7Jd and e).

7K
The trouble is, of course you, in a way, if you're a non-expert, you say 'Well, that's it.' The fact that it *is* written out in that way [as a picture], you say, 'Well, that's it, that's how it really is.' [Grant, 63]

7L
The important thing, of course, is not to kid the reader. If at the end of the day he winds up genuinely thinking that that's the arrangement of proteins, then that diagram has performed a *mis*-service . . . [The author must] warn the reader of this subjectivity. [Barton, 62]

7M
One danger that bothered me a little bit [in drawing picture VII], I don't know if I ever really expressed it too well, the fact that people would start taking it too literally. And some of my students did. But it was very few. That was because I gave it to them with the caveat, 'Hey, don't take this too seriously, it's a speculative view.' [Trubshaw, 97]

Many respondents, then, express awareness that pictures easily achieve a degree of facticity which, they are at pains to emphasise, is largely unwarranted. They maintain that pictures have a more powerful impact on most readers than do words or equations and they point to the 'danger' of their being 'taken as gospel'; particularly if they are specifically designed to create an illusion of realistic representation. In quotation 7M we see Trubshaw, one of the authors of the picture vii, addressing this issue in relation to his own students. Trubshaw had already stressed that one reason for making this picture more 'realistic' was to give students 'some semblance of reality to hold on to' (Trubshaw, 92). In 7M, however, he states that he was worried about students taking the picture *too* literally. Shortly afterwards in the interview, he mentioned that authors of several elementary biology textbooks had asked permission to reproduce this picture, and expressed a concern that the readers of these textbooks might also misinterpret it.

Trubshaw is faced with a dilemma over picture vii for the following reasons. First, like other scientists in his area, he treats pictures in these passages of reflexive verbal discourse as convenient fictions. He draws attention to their fictional components and stresses that his own picture is only ' "more realistic", in quotes'. Secondly, he maintains that many lay persons and students will be inclined to approach pictures from an empiricist or realist perspective. Thus, thirdly, he accepts that pictures intended for *this* audience will probably be more effective if they are presented in a style which can be easily read as being fairly realistic. But, at the same time, he stresses that realistic conventions must be used in a way which is not inconsistent with the fictional aspects of the pictures and which does not mislead non-specialists into taking the visual product *too* literally. Given their account of the fictional aspects of pictures and of the realist inclinations of outsiders, Trubshaw and his colleagues are faced with a precarious interpretative tightrope.

We will call this interpretative problem 'Trubshaw's dilemma'. The dilemma grows out of scientists' use of both realist and fictional repertoires to characterise their pictures. It arises out of the attempt to say that pictures are on the whole conventional fictions, yet that in certain contexts they can and should be designed so that they can be interpreted realistically. In this sense, Trubshaw's dilemma resembles the interpretative dilemma which we discussed in our examination of the TWOD. For the TWOD was a solution to parallel interpretative difficulties encountered as participants sought to combine the empiricist and contingent repertoires. However, the kind of dynamic resolution provided by the TWOD cannot easily be employed to solve the interpretative problem of apparently realistic representations which, at the same point in time, should convey their underlying fictional character.

It appears then, that Trubshaw's dilemma is by no means easy to resolve, either visually or verbally. Consequently, most of our respondents focus on the danger of encouraging over-realistic interpretation in pictures like VII and VIII, without going on to suggest any way in which it can be or is avoided in practice (see 7E, 7H, 7J, 7K and 7L). Trubshaw himself, who treats picture VII in more personal terms than most other speakers, resolves the problem in his account of his own students' actions, by claiming that he was able to guide them orally towards a correct reading (7M). But this hardly resolves the original dilemma. In the first place, it would seem that at best Trubshaw has reinterpreted his picture to his students in fictional terms and abandoned his claim that fairly realistic pictures are somehow appropriate for such an audience. In addition the broader problem, namely, how to prevent the great majority of students, who did not have the benefit of his authoritative advice, from misinterpreting the picture, is simply ignored.

There does not appear, then, to be any generally available interpretative device by means of which speakers resolve Trubshaw's dilemma at the verbal level. One reason for this may be that the dilemma is not confined to the realm of verbal discourse. Thus if there existed widely established *pictorial* conventions for transferring pictures from one interpretative context to another and for encouraging appropriate contextualised 'readings', some of our respondents would presumably have mentioned them in their discussions of picture VII and in their treatment of Trubshaw's dilemma. In other words, the dilemma may not merely be generated by interpretative problems in verbal discourse, but may be a response to parallel interpretative problems in the domain of pictorial organisation.

If this is so, we would expect that Trubshaw's dilemma would reappear in the pictures themselves. We can, in fact, observe the existence of relevant interpretative tensions in pictures VII and VIII if we compare them with picture II or picture V. As we have noted before, the pictorial style of the two latter pictures from the research literature is coherent, restricted and highly conventionalised. The main pictorial components are arrowed lines crossing a membrane consisting of parallel straight lines plus, in V, a circle for the ATPase. These conventional techniques of representation are also to be found in the two *Scientific American* pictures. However, in the latter two pictures other phenomena are represented in a way which departs significantly from the normal conventions. The membrane, for instance, is given a distinctly spongy texture. The impression is conveyed pictorially, by means of a visual metaphor with everyday objects, that the membrane is a softly resistant, mattress-like strip of material. Similarly, the ATPase projecting inside the membrane is not a conventional circle, but a bulbous and very specifically articulated knob. Perhaps most striking of all these 'recognisable' representations is the cog-like mechanism which is shown in picture VIII as being responsible for rotating the flagellum of the bacterium *E. coli.*

Owing to the non-reflexivity of the pictorial language used in these pictures, it is impossible for us or for participants to identify and discuss their organisation except in verbal terms. Nevertheless, it is possible to describe verbally the pictorial means used by Trubshaw and Smith to make their pictures 'more realistic'. They do this, we suggest, by combining the geometrical representations of the research literature with drawings which depict more naturalistically some of the objects involved in 'ox phos' and by making these objects resemble various kinds of recognisable everyday objects. There can be no doubt, of course, that these latter depictions are also conventional. But they convey a greater impression of 'realism' by the use of a representational style which is closer to that of everyday naturalism and by the visual metaphors with objects from the everyday world.

Trubshaw's verbally formulated dilemma, then, is reflected in the pictorial tension between these two kinds of representation in pictures VII and VIII. The dilemma can now be reformulated in more pictorially relevant terms as: 'Does the use of "realistic" components in addition to "conventional" ones encourage interpretations in realistic terms on the part of non-specialists? And if so, how can this be avoided?'

This restatement of the participants' interpretative problem reveals two interesting questions for us as analysts. First, 'Are the "realistic" components of pictures like VII and VIII ever interpreted as realistically as our respondents suggest? Do they actually provide pictorial resources which can be, in our respondents' terms, "misinterpreted"?' Secondly, 'Is there any evidence of scientists responding in *pictorial* terms to Trubshaw's dilemma?' We will explore these questions in the next two sections. Positive answers to either question would tend to indicate that we have been right to maintain in this section that Trubshaw's verbal dilemma has a genuine pictorial counterpart.

The flagellar motor and the Supreme Being

It is clear that any given picture can be interpreted in various ways by different viewers, in the same way that any verbal text is open to different readings. One reason for this is that the interpretation formulated by a viewer will be related to the kind of interpretative work in which he is engaged. Thus, when we refer to an interpretation of a picture (or to a reading of a text), we are not referring to uncontextualised processes occurring in an observer's 'mind', but to the way in which that observer uses the organisational features of the picture to construct a contextually relevant verbal and/or pictorial formulation of his own. In the next few pages we will examine one strongly realistic or empiricist interpretation of picture VIII by a non-bioenergeticist. Our goal will be to discern which components make possible the construction of an empiricist text by this non-specialist interpreter. We will show that, as our respondents 'feared', the non-specialist interpreter is able to extract from picture VIII certain apparently realistic elements and to treat them as directly observable phenomena in the real world.

Picture VIII, as we have noted, represents the membrane of the bacterium *E. coli*. Much of this picture closely resembles picture VII and the biochemical processes said to operate in this bacterium and in mitochondria are very similar. But at the top of the picture a unique feature of bacteria is depicted, the junction of the bacterium's flagellum (a kind of tail which is responsible for the organism's motion) with the membrane. The legend to picture VIII contains the following reference to the operation of the flagellum: 'The rotation of the flagellum is also powered by the influx of protons. At the root of the flagellum is a ring of 16 proteins, opposed to a similar ring in the cell wall. If a proton must pass through each protein to rotate the flagellum a sixteenth of a turn, 256 protons would be consumed in each revolution.' Unlike picture VII whose legend contains the warning, 'The arrangement of the molecules, however, is not yet certain, and the model presented here is somewhat conjectural', picture VIII is presented with no explicit indication that its realistic structures are not to be taken at face value.

Those of our respondents who were specialists on bacteria commented rather critically on this picture. For example:

7N
I do visualise things, but [this is] almost like an illustrated version of Jane Austen to me. I don't visualise the flagellum *quite* like that. I have my own image of it. [Roberts, 42]

7P
There isn't any evidence that that's a transmembrane protein. The same for these antiporters here. This whole depiction of the flagellum is, that's a real working conceptual hallucination . . . The sub-unit structure of this is highly hallucinatory . . . There *is* evidence for 16 sub-units and he's drawn them all as nice circles. It's just that it's a science fiction model of the flagellar rotor that's all. It does incorporate what's known. You can count them. [*Counts sub-units in picture* VIII.] Well, it's close to 16. So he's got 16 sub-units there. Nobody knows if they're really oriented that way. He's taken a certain poetic licence in doing it. That's OK. [Cookson, 81]

Despite these criticisms of the style and content of picture VIII, both respondents deemed the *Scientific American* article to be appropriate for its audience.

The non-specialist's response to picture VIII with which we are concerned appeared in the year following the publication of the *Scientific American* article. It is to be found in *Back to Godhead: The Magazine of the Hare Krishna Movement* and consists of an article entitled 'The machinery of evolution: OUT OF GEAR? A mathematician finds flaws in one of Darwin's basic assumptions'. It was written by Hardy, a mathematician from an American Ivy League college who is described as specialising in probability theory and statistical mechanics.[5] This article is of interest to us because one of its two pictures is a reproduction of part of picture VIII and because a discussion of the flagellar motor in the light of this picture is the fulcrum of an argument for the existence of 'a primordial, absolute personality' or 'Supreme Person'. It constitutes, therefore, a fascinating reading of picture VIII by a scientifically informed outsider. Despite the transcendental conclusion to Hardy's article, his text is formally organised in strongly empiricist terms and the existence of the Supreme Being is treated as an 'experimentally verifiable' phenomenon. We will give a short sketch of his argument before commenting on it.

Hardy begins with an idea which, he says, is essential to Darwin's theory of biological evolution. He formulates this idea as the hypothesis 'that the physical structures of all living organisms can transform from one to another through a series of small modifications, *without departing from the realm of potentially useful forms*' [emphasis added]. Hardy challenges the assumption that biological diversity could have occurred by means of evolution through a continuous series of useful forms. '[I]f there exist any significant structures in living organisms that *cannot* have developed in this way, then for these structures, at least, the hypothesis of evolution is ruled out, and some other explanation of their origin must be sought.' The flagellar motor of *E. coli*, as depicted in picture VIII, is taken to be a prime example of such a structure. A crucial step in the argument is that of proposing a close analogy between biological structures and mechanical structures. A picture is provided showing collections of cogwheels, gears, shafts, and so on. Most of these groupings of mechanical parts appear not to be organised in any systematic manner, but here and there, particular collections can be seen to function as organised units. 'If we visualise the space of mechanical forms, we can see that some regions in this space will correspond to wrist-watches and other familiar devices, and some regions will correspond to machines that are unfamiliar, but that might function usefully in some situation. However, the space will consist mostly of combinations of parts that are useful as paperweights at best.'

This representation of the 'space of mechanical forms' is said to be directly comparable to that of biological structures. 'These mechanical parts are comparable to the molecules making up the organs of the bodies of living beings. Since mechanical parts and molecules alike fit together in very limited and specific ways, a study of mechanical combinations should throw some light on the nature of organic forms.' By means of this analogy, Hardy displaces his spatial representation of shafts, levers and gears to the realm of biological organisms. Thus he is able to present us with a verbal and visual picture, in which: 'The class of all possible forms made from organic chemicals is like an ocean of tiny mechanical devices, most of them useless. The few useful forms are like islands surrounded by vast expanses of useless ones.'

Given this picture of the distribution of mechanical, and therefore biological, forms, Hardy concludes that it is impossible to maintain the Darwinian conception of evolution by means of gradual movements through a continuous series of useful structures. Movement from one 'island' to another cannot be gradual and continuous, because the intermediate forms are not viable, self-maintaining structures. He concludes that gradual structural evolution is impossible and that we must seek an explanation of the diversity of biological forms which recognises the need for radical leaps between complex biological systems.

Hardy formulates his own alternative to Darwinian evolution by introducing another analogy; this time between the discrete structures of the biological realm and the 'products of human creativity'. The latter, he suggests, often occur as sudden intuitive insights. Hardy then proceeds to argue that: 'If it is the nature of biological form and the forms of human invention to exist as isolated islands in the sea of possible forms, then some causal agency must exist that can select such forms directly. The experience of inventors indicates that this agency lies outside the realm of human consciousness or control . . .' From this conclusion, it is but a small step to argue that both biological forms and the products of human creativity are the intentional outcomes of a higher or more inclusive personal agent. The *Bhagavad-gita* is then quoted as having clearly identified the 'primordial, absolute personality', the Supreme Being, ultimately responsible for all organised systems. He ends the article with some suggestions for improving the investigatory procedures of science: a personal avenue of approach to the knowledge held by the Supreme Being, he suggests, already exists in the methods of bhakti-yoga which are similar to those of modern science in that both depend on 'clearly specified procedures leading to reproducible results'.

Clearly, this article is rich in topics worthy of investigation, but we will concentrate on examining how its author has interpreted and used picture VIII. The first point to note is that Hardy interprets this picture and its original text in exactly the way that our respondents said was expected of readers of such popularising outlets as *Scientific American*; that is, he extracts the central ideas rather than the precise details.

7Q
The motors are presently thought to be driven by a flux of protons flowing into the cell. Each motor is thought to consist of a ring of sixteen protein molecules attached to an axle . . . Although the exact details of the *Escherichia coli*'s molecular motors have not been worked out, *we can see that* they depend on the precise and simultaneous adjustment of many variables. In the space of possible molecular structures, the functional motors will represent a tiny, isolated island [emphasis added].

In this passage, Hardy acknowledges the uncertainties in current knowledge about the flagellar motor. He stresses that scientists *think* that the motor operates in a certain way, but that the details have *yet to be worked out*. Thus in his verbal discourse the precise molecular mechanism is treated as somewhat speculative. All that is taken as firmly established is the general idea that flagellar activity is somehow powered by the movement of protons across the membrane. But Hardy's overall argument cannot be developed further without a specific organic device which can be seen to resemble a wristwatch in its finely articulated structure.

This need is satisfied by Hardy's reproduction of the top part of picture VIII. His argument is carried forward, as we can see in quotation 7Q, by in effect *pointing* at the picture and by treating it as a literal representation of part of a functioning organism. Despite the *verbal* references to uncertainty, Hardy states that we can actually *see* the precise adjustment of mechanical parts within the structure of *E. coli* in his version of picture VIII. The recognition of uncertainty in the verbal text and the implicit admission that the picture must be to some unknown degree speculative are effectively erased by the apparent facticity of the pictorial representation. Thus this outsider's use of the flagellar motor as a crucial scientific datum for his argument is made possible, not by current experimental findings on proton translocation in bacteria, but by the fact that Trubshaw and Smith chose to present the general idea of the motor in visual form; by the fact that they adopted a 'realistic' or 'mechanistic' visual idiom for this pictorial component and by the fact that Hardy was thereby able to treat what the specialists described to us as at best speculative and at worst an hallucination, as an observable fact about the organic world.

The example of the flagellar motor and the Supreme Being is undoubtedly unusual in certain respects. Few readers of scientific pictures will use them to reach explicitly transcendental conclusions. Yet these unusual features may well be superficial. For Hardy's reading is formally a strongly empiricist one and in that important respect it may be typical of much of the interpretative work carried out by non-specialists within the broader scientific community. If this example *is* at all representative, it appears that pictures and the knowlege-claims which they embody may sometimes undergo reinterpretation and transformation of meaning as they cross the boundary between research networks and their members' wider audience. Furthermore, it seems that researchers may on occasion describe this transformation of meaning as one of *mis*interpretation; even though their own interpretative practices in some cases actively foster the kind of reinterpretation to which they object.

We have exemplified this process with a reading by a scientist who is fully trained and presumably technically competent in his own field. We are led to wonder, therefore, how frequently it is likely to occur among students. We are thus led back to Trubshaw's dilemma; that is, 'How can scientists use the powerful impact of pictorial representation to communicate with students educated in an empiricist tradition, without leading students to take the pictures "*too* literally"?' If this is a recurrent interpretative dilemma facing scientists, arising as we have suggested out of structured variations in social context and in forms of discourse in science, we might expect that scientists themselves will have sought ways of devising forms of pictorial organisation which begin to resolve it. In the next section, we will examine one interpretative device for doing exactly that.

Visual jokes and degrees of realism

At the beginning of this chapter, we presented as our first picture a representation of a joint animal–plant cell which was taken from a student textbook. One of the regular features of this book, and of others published for a student audience, is its use of humorous pictures or visual jokes. There are several such pictures per chapter in the book and each chapter ends with a visual joke which summarises its contents. These jokes are accompanied by a much larger number of drawings of technical apparatus along with obviously schematic representations of cellular phenomena and more realistic-looking pictures. The first chapter of 20 pages contains three visual jokes and about 21 other pictures.

This illustrates the considerable use of visual discourse which is often found in student textbooks. But what contribution do visual jokes make to such discourse? We will try to indicate one of the things which can be done with visual jokes by looking at and commenting on the joke picture which concludes the chapter on the creation of proton gradients across mitochondrial membranes.[6]

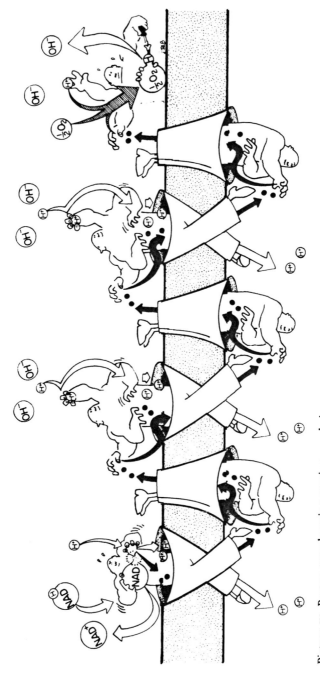

Picture ix. Proton-translocating respiratory chain

This picture is seen as a joke because it has certain stylistic and formal properties. (We will examine the structure of scientific jokes in more detail in the next chapter.) Formally, like a great many jokes both scientific and otherwise, it combines elements from two normally quite separate areas of discourse. The first class of elements in this case consists of the various standard scientific symbols and the pictorial representation of the mitochondrial membrane. The two parallel horizontal lines are recognisable as a membrane, partly by their position at the end of a chapter on the movement of ions across membranes and partly by their spatial relationship with the symbols representing protons, oxygen and the other constituents specifically required for proton translocation in mitochondria. The symbols NADH, OH^-, H^+, etc., are placed visually in relation to the membrane roughly as the preceding text describes the location of the corresponding phenomena in mitochondria. This confirms for the reader that the picture in some sense represents the mitochondrial membrane as well as the processes whereby proton gradients are created across the membrane.

These elements of biochemical discourse, however, are combined in picture IX with representations which are immediately recognisable, despite their distortions, as human figures wearing voluminous trousers. Thus, the incongruity essential to humour is created by the juxtaposition in the picture of elements of discourse, both verbal and visual, which are difficult to reconcile in a literal sense. The humorous intent of the picture is also made evident by the multiple arms and faces, neckless heads, identical physiognomies and extended trousers which are reminiscent of comic-book characters.

Picture IX, like the other end-of-chapter pictures in this textbook, is organised in a way which tells the reader immediately 'not to take it too seriously'. Nevertheless, it is linked directly to the preceding text and constitutes a summary account of what is claimed in that text to be known about the formation of proton gradients in mitochondria. In this respect it closely resembles other pictures we have examined which deal with the mitochondrial membrane (III, IV, V, VII and VIII), and like them, it uses the normal research conventions of two parallel straight lines crossed by proton-carrying arrows to convey the non-humorous content in a stylised, fictional way. But, unlike pictures VII and VIII, which create the impression that the picture represents the real biochemical world by embedding these conventional elements in a more 'realistic' environment, picture IX contrasts them with even more obvious fictions, namely, the baggy-trousered mannikins. In this way, we suggest, picture IX, although overwhelmingly fictional in its import, is organised in a manner which attaches different *degrees of realism* to its varied components. As a result, the formal structure of picture IX more closely resembles the verbal interpretations furnished by our respondents, whilst at the same time retaining the capacity to convey a basic scientific interpretation of the processes of ATP production.

In the textbook chapter which precedes picture IX, the discussion of the mechanisms of proton translocation begins with the sentence: 'Two major types of hypotheses have been advanced to explain how this proton pumping is achieved.'[7] The text then proceeds with an exposition of both hypotheses including two corresponding schematic representations in the style of picture II, with no attempt being made to choose between them. The mechanisms of proton pumping, then, are treated in the text as an area of uncertainty. It is clearly difficult, if not impossible, to cope with uncertainty within a single pictorial representation; at least, as long as some attempt is made actually to re-present the phenomena in question. But in the summary picture at the end of the chapter, the device of the visual joke is used in a way which surmounts this difficulty. For it is precisely the speculative mechanisms of proton pumping that are represented by the obviously fictional man-like figures. The effect of this is to ensure that the summary picture confirms those aspects of proton translocation which in the text are taken as established, whilst humorously reminding the reader that certain scientific questions remain unanswered. It indicates that the phenomena being represented have varying claims to be regarded as scientifically accurate, without departing from a strongly fictional overall perspective and, thereby, without falling foul of Trubshaw's dilemma.[8]

It appears, then, that the visual joke in picture IX avoids Trubshaw's dilemma by introducing reflexivity. Its organisation contains a comment on the nature of its own discourse and an indication of the degree to which its constituents are to be taken seriously. The importance of this reflexive structure is that it furnishes a pictorial solution to Trubshaw's dilemma. In this sense, picture IX brings the present analysis full circle. We began this chapter by emphasising the importance of pictorial discourse in science and the crucial part it plays in transferring scientific knowledge from the research community to non-specialists. We then identified a series of formal properties of a particular class of pictures and we suggested that these properties seemed to be more closely aligned with the characteristics of the contingent verbal repertoire than with those of the empiricist repertoire. At this point, we examined scientists' verbal accounts of pictorial discourse and we found that, although they talked about these representational pictures in both realist and fictional terms, their characterisations were overwhelmingly fictional. We suggested that there are strong parallels between fictional and contingent discourse about pictures, and between realist and empiricist discourse about action; and that respondents' verbal portrayals of pictorial discourse in our interviews closely resemble our own specification of the formal properties of this discourse.

The issue of communication with non-specialists was approached through participants' talk about the contextual character of their pictures and about the need to provide a greater degree of realism for certain kinds of outside audience. These verbal reflections by participants gave rise to the interpretative problem of Trubshaw's dilemma which, we suggested, has a direct counterpart in the pictorial organisation of some popular pictures. Through a close examination of a particular text, we showed that the 'realistic' components in such pictures could facilitate strongly empiricist interpretations and that non-specialists could treat such components as literal representations of the real world.

Finally, we saw that visual jokes can supply a pictorial device for expressing scientific uncertainty and, thereby, a way of organising pictures to convey 'degrees of realism'. The combination of unlikely visual elements can provide the structural basis for a joke; whilst at the same time resolving Trubshaw's dilemma by conveying the scientific accuracy of the main established ideas without running the risk of too literal an interpretation of speculative conceptions. The fact that scientists sometimes appear to deal with Trubshaw's dilemma by means of a specific pictorial device provides some indication that our prior analysis of the nature of pictorial discourse and of the interpretative problems which arise in the course of communication across interpretative boundaries is 'along the right lines'.

8

••

Joking apart

In this chapter, we will focus on scientists' humour and on its connections with other elements of scientists' discourse. We will pursue the joint aim of extending the scope of our analysis to cover a component of scientific culture which has largely escaped the attention of sociologists, and of using this analysis to furnish a check on some of our earlier conclusions. In the examination of scientific humour which follows, we will discuss examples from the scientific research 'community' at large. As a result, we will be able to show that our previous analysis, although based on a limited range of data obtained from a single research network, provides new insight into discourse generated by scientists in quite different fields.

Humour as a sociological topic

Participants' published reflections about the existence of humour in science frequently emphasise that science is a very serious business, but that, nevertheless, science actually contains and should contain a strong humorous undercurrent. In such reflections, it seems to be taken for granted that non-scientists are largely unaware of the importance of humour in science.[1] It is perhaps necessary for us, therefore, to begin by making two similar points: first that, despite appearances to the contrary, there is a pronounced humorous element in scientific culture; and secondly, that scientific humour is a particularly significant site for sociological analysis.

The regular and organised production and dispersal of scientific humour can be easily documented. For instance, many research laboratories produce humorous magazines. The members of the Sir William Dunn Institute under Gowland Hopkins produced a comic journal[2] every year between 1923 and 1931. An earlier example, also from Cambridge, is the *Post-Prandial Proceedings of the Cavendish Physical Society* in the J. J. Thomson era. There are, in addition, three famous Festschrifts for Niels Bohr issued by the Institute of Theoretical Physics, Copenhagen, in 1935, 1945 and 1955, under the title *Journal of Jocular Physics*. In more recent times, scientific humour and irony have become institutionalised in several 'joke journals'; the *Journal of Irreproducible Results*, the *Worm Runners' Digest*, the *Journal of Insignificant Research*, the *Subterranean Sociology Newsletter* and the *Revues of Unclear Physics*. The first four of these are described and discussed in a review by Garfield.[3] Furthermore, several varied collections of scientists' humour have been published[4] and a paperback containing cartoons from *American Scientist* is currently available to the general public.[5] There is also the widespread use of visual jokes in scientific textbooks, which we have already noted; and students in search of further light relief can now obtain a complete Introduction to Biochemistry written in verse and set to music.[6]

It is clear, then, that scientific culture is by no means entirely serious. But even if humour does occur regularly in all the main realms of cultural production in science, that is, in the laboratory, the research and teaching literature and the popularising media, it is still possible to question whether there is much to be gained by studying it. One might respond with the view that humour is unlikely to tell us anything interesting about the serious side of science and that the study of scientists' humour is, therefore, an analytical irrelevance; except in so far as humour is an interesting topic in its own right. However, we wish to advance an exactly opposite view. We suggest that scientists' humour, and indeed humour generally, is a crucial sociological topic[7] and that the study of humour can play a central part in putting our kind of sociological analysis to the test. Let us clarify this claim.

One of the major conclusions of the general literature on humour is summarised in the following quotation.

> Most of the theorists I have cited (as well as those not quoted here) agree, once allowance is made for different ways of putting things and different emphases, that a necessary ingredient of humour is that two (or more) incongruous ways of viewing something (a person, a sentence, a situation) be juxtaposed. In other words, for something to be funny, some unusual, inappropriate, or odd aspects of it must be perceived together and compared.[8]

This description of the basic structure of humorous discourse can be linked in a very simple and direct fashion to the analytical approach adopted in this book. We have proposed that scientists, like other social actors, regularly employ divergent repertoires to construct versions of their social world which often appear to be literally incompatible. In most circumstances, we have suggested, these repertoires are kept separate or are applied to distinct social categories. As a result, obvious interpretative inconsistency is kept to a minimum in ordinary, serious writing and conversation. However, inconsistency or incongruity seems to be a necessary, or at least a very frequent characteristic of humour. Much humour seems to depend on precisely the intimate juxtaposition of, and sudden movement between, divergent interpretative frameworks. We would expect, therefore, that, in constructing humorous incongruity, participants will often draw on recurrent interpretative repertoires which are normally kept apart.

It seems to follow that if our prior analysis of scientific repertoires is broadly correct, and if these repertoires are widely used among research scientists, we will find them regularly juxtaposed in scientists' humorous and ironic formulations, and their potential incompatibility emphasised. We suggest that this is, in fact, the case: humour constructed by scientists for other scientists is often accomplished by combining and contrasting the empiricist and contingent repertoires. Moreover, contextual differences between formal and informal discourse, and recurrent interpretative patterns such as that of asymmetrical accounting for error, are treated by participants as topics for ironic comment. In addition, Trubshaw's dilemma reappears in terms of verbal, as well as visual, humour. In short, the relevance of our preceding analysis to naturally occurring discourse in science is amply confirmed by the study of scientists' humour.

On several occasions in previous chapters we have already observed instances of scientists' humour which depend on the juxtaposition of normally discrete interpretative repertoires. In the last chapter, for example, we saw how elements from the lay world of comic strips were combined in picture IX with elements of technical scientific discourse to create humorous incongruity. A more subtle example occurred in chapter four and is reproduced again below.

> **4H**
> 1 There are lots of things you have to take into account. 2 And there are very strong individuals in the field who want to interpret everything in terms of their theories. 3 Of course, those are the other guys, not us. 4 We're interpreting it even, balanced [*general laughter*]. 5 The other ones are the ones who are doing that. 6 When you try and bend the data like that sometimes you don't take into account everything, too. [Hargreaves, 51]

In this passage, the speaker ironicises[9] his own account of others' errors by his tone of voice in sentences three to five. As he changes his style of delivery, he suddenly switches from a straightforward, internally consistent account of error in asymmetrical empiricist terms into a different interpretative framework. By his ironic tone of voice, the speaker draws attention to the possible partiality of his own account, thus acknowledging implicitly the existence of the recurrent pattern of asymmetrical accounting for error, and to the possibility that his own scientific claims could be taken to be as questionable as those he is criticising. As he juxtaposes and moves between the empiricist and contingent repertoires in this passage he generates laughter along with interpretative inconsistency. The ironic tone of voice enables the speaker to do this without any lexical or structural changes in his discourse by instructing the hearer to carry out the necessary alterations in meaning for himself.

This example illustrates how a scientist can construct humour informally by moving between the two repertoires identified in our earlier analysis. At the same time, the speaker's joke is made possible by an implicit recognition of the frequent occurrence of an interpretative pattern which we have called asymmetrical accounting for error. Thus this humorous passage provides indirect evidence of the regular appearance of that pattern in participants' discourse. It also indicates that humorous incongruity is likely to be disregarded and 'not taken seriously'. For not only does the speaker return without hestitation in sentence 6 to an empiricist account of others 'bending the data like that', but the interviewers also pay no further attention to the inconsistencies implicit in the speaker's ironic aside. Although when the interview tape was being transcribed we noted that sentences 3 to 5 were 'said ironically', at the time of the interview we simply joined in laughing at the suggestion that there might be competing versions of participants' actions and moved immediately on to other topics. There is an indication here, then, that humorous incongruity is more likely to be bracketed and treated as inconsequential than other forms of interpretative inconsistency.[10] In this respect, humour may resemble the TWOD in enabling speakers to move between repertoires and to draw attention to the existence of incompatible accounts, without thereby jeopardising their ability to reinstate one version over others and to treat the supposed coherence and consistency of ordinary serious discourse as unproblematic. If one focuses on the interpretative incongruities built into a humorous remark, the speaker can always respond with 'I was only joking'. This further strengthens the view adopted in this chapter that significant interpretative discontinuities, which are not easily visible in serious discourse, will often be revealed in humorous texts and informal jokes.

Interpretative repertoires and the proto-joke

We do not intend to offer here an analysts' definition of humour. We will instead take for granted that for participants there is a close family resemblance between verbal exchanges which lead to laughter, spontaneous witty remarks, formalised funny stories, written satire and cartoons. We will assume that participants use the term 'humour' to refer to all such discourse and we will identify particular instances by observing how they are categorised by participants. From this point on, we will concentrate on examples of humour which are widely available in written form and which are widely recognised as humorous in science.

Our first example, which has no storyline, no description of supposed events and no comic resolution, we refer to as the 'proto-joke'. We came across this proto-joke in the course of visiting biochemistry laboratories to carry out interviews. We first noticed it pinned on a laboratory notice-board. We smiled, but did not recognise its significance. Later, when its analytical importance became clearer to us, we began to ask biochemists whether they were familiar with it. Many of them were and we were able to obtain copies from four different research groups. They are all very similar and appear to derive from a book published in 1962,[11] which in turn makes use of material published several years earlier. The proto-joke consists of two lists of phrases, one referring to formulations which can be used in the formal research literature and the other supplying their informal equivalents. The lists are given titles like 'A Key to Scientific Research Literature' or 'Dictionary of Useful Research Phrases'. The version below reproduces some of the more popular items in our collection.

8 A

What he wrote	*What he meant*
(a) It has long been known that . . .	I haven't bothered to look up the reference.
(b) While it has not been possible to provide definite answers to these questions . . .	The experiment didn't work out, but I figured I could at least get a publication out of it.
(c) The W-PO system was chosen as especially suitable . . .	The fellow in the next lab had some already prepared.
(d) Three of the samples were chosen for detailed study . . .	The results on the others didn't make sense and were ignored.
(e) Accidentally strained during mounting . . .	Dropped on the floor.
(f) Handled with extreme care throughout the experiment . . .	Not dropped on the floor.
(g) Typical results are shown . . .	The best results are shown, i.e. those that fit the dogma.
(h) Agreement with the predicted curve is:	
Excellent	Fair
Good	Poor
Satisfactory	Doubtful
Fair	Imaginary

(i)	Correct within an order of magnitude . . .	Wrong.
(j)	Of great theoretical and practical importance . . .	Interesting to me.
(k)	It is suggested that . . . it is believed that . . . it appears that . . .	I think.
(l)	It is generally believed that . . .	A couple of other guys think so too.
(m)	The most reliable results are those obtained by Jones . . .	He was my graduate student.
(n)	Fascinating work . . .	Work by a member of our group.
(o)	Of doubtful significance . . .	Work by someone else.

When we discussed with biochemists lists like that reproduced as example 8A, we were told that they were meant to be humorous. One of our donors wrote of his list that 'It has been posted on the board in my office for some years now and many people have been amused by it.' We want to ask: 'What makes it funny? How is the humour produced?'

In the first place, it is clearly produced by drawing on our two interpretative repertoires. The phrases under 'What he wrote' are organised in terms of the empiricist repertoire and those under 'What he meant' in terms of the contingent repertoire. The empiricist phrases present scientists' research actions as impersonal (j to o), as following from procedural rules (c, d, g), and as allowing the facts to speak for themselves (c, d, f to i). But these phrases are translated into the informal idiom of scientific talk in a way which undermines their implicit conception of scientific action. Impersonality is replaced with personal commitment (b, d, g, j, k) and the influence of social relationships is stressed (c, l to 0). Similarly, the procedural rules of scientific investigation are depicted, not as determining scientists' judgements, but as being used by scientists to further their own knowledge-claims and their own interests (g to j, m to o). In general, the formal phrases of the research literature are reinterpreted to reveal the contingency of scientists' actions (c, e, f) and the contingency of their claims about the natural world (a, b, d, g to o).

This form of scientific proto-joke, then, achieves incongruity by contrasting repertoires in the most direct fashion. The long list of phrases, the continuation dots at the end of each formal phrase and the absence of any specific speaker all tell us that these are not isolated, idiosyncratic turns of speech, but that they represent distinct, coherent linguistic styles. The proto-joke is textually organised as a joke about interpretative repertoires. In scientists' informal talk, as we have seen, these repertoires are usually kept separate or they are reconciled by the use of appropriate interpretative devices. In this joke, however, they are brought together in a way which emphasises their incompatibility. We are told that this is what scientists *wrote* about their actions when they were constructing research reports, but it is not what they *meant*. The phrases of the contingent repertoire are presented as if they conveyed a literal description of what really happened. Incongruity is thus created in a way which favours the contingent perspective. As a result, it is the empiricist version of scientific action which is made to appear inappropriate and misleading. The text is organised in a manner which makes us laugh or smile at the foolishness or hypocrisy of the formal literature. The proto-joke comes close to being a satire directed at the official discourse of science.

The existence of various kinds of interpretative repertoire in science receives further support from the frequency with which scientists' jokes are constructed along the same lines as the proto-joke. There is, for example, the 'Do-it-yourself *CERN* Courier writing kit' which provides the reader with batches of numbered standardised phrases dealing with topics of relevance to *CERN*.[12] Readers are recommended to take 'any four-digit number . . . and compose your statement by selecting the corresponding phrases from the following tables . . .' There are numerous jokes which deal in this fashion with the existence of localised or contextualised repertoires. Of most direct interest to us here is the 'Conference Glossary' first published in 1960.[13] This is very close to the proto-joke in consisting simply of two lists of phrases likely to be heard in formal conference sessions. The lists are headed 'When they say', and 'They mean'. Their content is very similar to that of 8A. For instance, 'We have a tentative explanation' is translated as meaning 'I picked this up in a bull session last night.' This is a further indication that the two forms of accounting identified in our analysis do appear in naturally occurring interaction among scientists.

Sociologists of science have been inclined in recent years to organise their own analysis of scientific action in a manner similar to that found in 8A; that is, they have used scientists' contingent accounts as evidence for their own contingent interpretations of social action and belief. In contrast, philosophers have been more inclined to treat the formal literature as primary and to discount the contingent element in scientific action. We wish to emphasise that scientists themselves can and do adopt either perspective, depending on the kind of interpretative work in which they are engaged. Thus scientists employ the empiricist repertoire when writing research papers. When justifying their own scientific views informally they may use both repertoires, but they will usually treat the empiricist repertoire as primary. The same individuals, however, may favour the contingent repertoire when providing an explanation of false belief, when describing laboratory practice, when making a joke or when satirising the research literature. Neither repertoire provides the analyst with literal descriptions of events occurring in the research community. Rather the two repertoires must be seen as interpretative resources which are used by scientists to construct versions of events appropriate to varying interpretative contexts.

This section has shown that our previous analyses of scientific discourse provide insight into the way in which scientific humour is organised. Example 8A depends on the existence of divergent repertoires, on the gap between the formal literature and informal talk about science, and on the possibility of generating humorous incongruity through the characterisation of given actions in incompatible ways. Let us now move on to examine another kind of humorous product, the satirical research article.

Satire as a form of scientific discourse

The *Journal of Irreproducible Results* is described on its cover as a 'satire of interest to professional and scientific workers'. Many of the contributions to the *JIR* appear to make much use of irony, that is, they use words to convey the opposite of their usual meaning. In the course of conversations, irony can be conveyed by means of gesture, emphasis or tone of voice, as it was in 4H. In written texts, however, this is not possible and various kinds of clues are normally built into the textual structure to indicate that it should not be accepted at face value. This can be observed in example 8B, which is the full text of an article in the *JIR*.

> **8B** *Dose-Response Curve to Oral Lactose in Spontaneous Hypertensive Rats*
>
> *Abstract*
> 1 Goodman and Gillman have not included lactose in their 8th revised edition of *The Pharmacological Basis of Therapeutics*. 2 In other well known reference sources we found the same dearth of information. 3 Following a computer library search we failed to find any study where a lactose dose-ranging study was reported, which is standard result for a computer search. 4 This paper reports the effect of 'no-treatment' in comparison with that of 0.5, 5 and 50 mg/kg of lactose on the mean arterial pressure (MAP), and heart rate (HR) of slanted-eyed spontaneously hypertensive rats (SHR).
>
> *Materials and methods*
> 5 Male rats of 290 to 350 g body weight and 30 to 35 weeks of age were purchased from Rats Incorporated, Wilmington, Delaware. 6 The

modified* tail cuff method was used to indirectly record the systolic blood pressure. 7 Twenty-four rats (6 per group) were placed in restrainers and kept at room temperature (27 °C) and had free access to water ad libido. 8 The systolic blood pressure from each rat tail cuff was recorded six at a time on a Hewlett-Packard 7754A four-channel recorder coupled to a Sansui Quadraphonic Receiver. 9 Control readings were made hourly for four hours at the same time of the day for four days before treatments. 10 Following oral administration of the lactose solutions (group II, III, IV) or no treatment (group I), the same recordings of the MAP and HR were made on the four days of treatment. 11 A one-day post-treatment period was also taken to establish if recovery had occurred.

Results

12 The effect of lactose on systolic pressure in groups of six rats measured indirectly (tail plethysmography), indicated that no dose–response relationship was obtained but lactose at 50 mg/kg, (group IV) significantly lowered ($p<0.05$) the systolic pressure by an average of 3 mm Hg and significantly increased the HR ($p<0.05$) in the same group, by an average of 4 beats per minute. 13 The effect occurred 1 and 2 hours postdrug on Day 1 but disappeared thenceforth. 14 MAP and HR were back to control levels in all groups from the second hour on and remained at approximately the same level during the post-treatment day.

Discussion

15 The indirect tail cuff (tail plethysmography) has been reported to overestimate systolic blood pressure. (Personal communication.) 16 If this is the case in our experiment, we must give consideration to the possibility that lactose may be more hypotensive than our data would suggest.

17 In order to verify this we are presently conducting a similar experiment using the direct (aortic cannula) method.

18 In conclusion, we state that 50 mg/kg of lactose p.o. appears to have a short depressing effect on the MAP in SH rats and we urge all those who have used this treatment as a placebo in their study of hypertension to reconsider the validity of their data.

*The cuff lining is made of cat fur to add to the stress (manufacturer's brochure).

Let us describe some of the structural features of this text. In order to avoid being unnecessarily clumsy, we will sometimes use conventional phrases like 'the author provides' or 'the author is certain that'. Such phrases should not be read, however, as referring to the author's intentions, but as referring to features of the text. Thus, to assert that 'the author provides indications of humorous intent' is not intended as a claim about the author's actual intentions, but as a statement about certain interpretative elements in the text. Nevertheless, we did obtain brief comments on our analysis from the author. These will be mentioned parenthetically as author's readings of the text.

We suggest that the author provides clear indications of humorous intent early in the paper. The second part of sentence three, suggesting that computer library searches are usually worthless, although acceptable as an informal comment, would not normally be found in the formal literature. The early introduction of this phrase from informal discourse quickly generates a modest incongruity.

[Author's reading: The paper was written somewhat in jest at a time I was with Merck & Co. Annoyed at being given second-rate projects to work with which did not lead to publication I decided to write the *JIR* paper: at least I would have a publication that year! Furthermore, I wanted to show some of my colleagues with a high opinion of themselves that everything had not been said about hypertension. There is also a little dart at literature search: I have always done better by myself than the computer.]

A second clue that the text is ironic is provided towards the end of sentence four, where the 'spontaneous hypertensive rats' of the title are described as 'slanted-eyed'. Nevertheless, this 'clue' is by no means unambiguous for the non-specialist. There is nothing in the text to tell those who have no dealings with experimental rats whether or not such slanted-eyed creatures exist or whether, if they do exist, they have well-known experimental advantages.

[Author's reading: The 'slanted-eyed rats' statement was 'my way of referring to a Japanese strain of rats known as the Okamoto-Aoki strain which has been bred to develop spontaneous hypertension'.]

Perhaps a more significant feature of the paper's abstract is the fact that there is no mention at all of the actual experimental findings. A brief summary of the results is, of course, a normal feature of abstracts in the serious literature. The ironic implications of its absence become clear only after we have read the whole paper and have realised that there are no findings.

The methods and materials section is the most explicitly humorous part of the article. The picture is conveyed of these groups of six slanted-eyed rodents being plugged by their tails into what reads like a hi-fi system, with cat fur being added to the equipment to increase the stress. The impression of incongruity is strengthened when it is noted that stress would tend to raise the animals' blood pressure and thereby interfere in an uncontrolled manner with the observed effects of lactose on MAP. Thus the materials and methods section creates two kinds of incongruity. Not only are various details of the experiment bizarre, but, even if such details are ignored, the overall research design seems to be scientifically flawed.

In the results and the discussion sections, the humorous content is less obvious. However, sufficient clues of humorous intent have already been provided to encourage us to look for a double meaning behind the empiricist terminology of these sections. The data are presented and discussed in the apparently straightforward terms of the formal literature and with no further attempt to convey comic images. Yet these two sections are the most firmly ironic of the whole text, for there are actually no results; certainly insufficient results to add to existing knowledge of the physiological effects of lactose. Thus the paper becomes a satire on how scientists try to present trivial observations as if they were scientifically valuable.

The basis in the text for this conclusion can be seen most clearly by comparing the title, which proclaims that the paper contains a 'dose-response curve to oral lactose', with sentence 12, which states explicitly that 'no dose–response relationship was obtained'. The 'results' obtained in this imaginary experiment can be summarised in the statement that, over a period of nine days, the blood pressure and heart rate of one group of animals departed from the average for a period of up to two hours immediately after the start of the experimental treatment. There is, therefore, no dose–response curve at all, but merely a single, isolated departure from the normal. The author claims to be 95 per cent certain that this fluctuation was not due to chance. But whether it is replicable and whether it is a side-effect of the experimental situation are left quite unclear.

Three of the four sentences of the discussion, instead of offering some tentative explanation of positive results as is normal in Discussion sections, focus on the *failure* to produce any experimental evidence of the supposed relationship between blood pressure and oral lactose. Sentence 15 makes use of a 'personal communication' to suggest that the technique of measurement employed in the experiment systematically under-represented the real empirical effect of lactose. The author uses this communication, which of course cannot be checked by the reader or by the referee, to reaffirm the claim embodied in the title of his paper. In sentence 16 he maintains that, if his measurement technique *was* defective, the relative insignificance of the findings is perhaps misleading. In other words, the failure to obtain positive results *with this technique* has now been made to appear to be exactly what one would expect. Another effect of the personal communication is to make this experiment worthless, because it reveals that the techniques employed do not properly measure the variables they are supposed to measure. But the author does not develop this implication. Instead, in sentence 17, he uses his initial experimental 'failure' as a basis for justifying a new study, the purpose of which is to describe that dose–response curve which was to be the subject of this article.

The final sections of this paper, then, reveal it to be an ironic gloss on the second pair of phrases given in 8A: namely, 'While it has not been possible to provide definite answers to these questions . . .' and 'The experiment didn't work out, but I figured I could at least get a publication out of it.' There is also a pronounced reference to 8Ah: 'Agreement with the predicted curve is fair/imaginary.' Thus this text can be interpreted as an elaboration on specific components from the two repertoires discussed above. This is not immediately obvious from a first reading. But we can now see it as a strongly ironic text which is organised, like the proto-joke, to bring out the supposedly contingent character of many serious scientific knowledge-claims and to reveal how scientists can employ the formal mode of discourse to attempt to hide this contingency.

Because the author of this paper has to provide indications of ironic intent in his text, this spoof paper can do no more than resemble proper research articles in certain formal respects. It is implied, nevertheless, that the construction of serious papers in the formal literature differs from that practised here only in the sense that interpretative clues will not be emphasised and the deceptive hiding away of contingency will be that much more effective.

[Author's readings: 'Your perception of my article is accurate . . . Your article is more universal than mine.']

It is interesting to note that in both the proto-joke and in this ironic article, the contingent repertoire is treated as primary. In 8B, of course, the contingent repertoire is no more than hinted at in the text. Nevertheless, it is possible to read the text as ironic only by contrasting the ostensibly empiricist version of actions and conclusions which it offers in the last two sections with some informal account of 'what really happened in the lab and what the results really mean'. Thus, in order to display the text as ironic, we were obliged to offer above an informal account of 'what the results actually were', which was based on the content and structure of the text, but which was nowhere fully explicit in that text.

Some kind of extraction of an alternative version from the text is unavoidable in cases of irony, because the point of irony is to keep largely implicit the juxtaposition of those incongruous perspectives, repertoires or versions of events on which the humour depends. We have tried to show that the article above is organised to suggest an alternative version of the author's actions and results in the terms of the contingent repertoire. We have also tried to show that it is a concrete exemplification of one of the themes contained in the proto-joke; and that its ironic structure can be understood in relation to the two repertoires of discourse embodied in the proto-joke.

Scientific and non-scientific humour

We suggested at the beginning of this chapter that humour is a critical topic, not just for the sociology of science, but generally for any systematic sociological analysis of discourse. In this section, we will discuss a joke taken from an area of social life far removed from the research community. Our aim is to show that the relationship which we have observed in our material between recurrent interpretative patterns and the structure of humour is not unique to science or to our collection of data.

The joke with which we are concerned is as follows:

> 8 C
> *Doctor:* I've got good news for you, Mrs Brown.
> *Patient:* It's Miss Brown, actually.
> *Doctor:* I've got bad news for you, Miss Brown.

This joke is taken from MacIntyre's analysis of what she calls, following Mills, the vocabularies of motives for dealing with pregnancy in modern Britain.[14] She suggests that there are two main vocabularies in use, two 'versions of reality' as she puts it at one point, each of which is linked to a distinct social category. She shows how doctors, nurses, and social workers, whilst regularly professing that reproduction is 'natural, normal, and instinctive' for women generally, systematically provide 'different sorts of accounts' for the responses and motives of unmarried as compared with married women. There is, then, considerable similarity between, on the one hand, our identification of two interpretative repertoires in science and their application to different social categories in accounts of error and, on the other hand, MacIntyre's analysis of accounts of pregnancy.

The experience of pregnancy for persons defined as 'married woman' is given meaning in terms of a repertoire with the following main features:

(1) Pregnancy and childbearing are normal and desirable, and conversely a desire not to have children is aberrant and in need of explanation.
(2) Pregnancy and childbearing are not problematic, and to treat them as such indicates that something is wrong.
(3) Legitimate children with a living parent should not be surrendered for adoption or taken from the mother, as this would occasion too much distress for the mother.
(4) If a couple is childless it is clinically advisable that they receive diagnostic attention and, if necessary, treatment for infertility.
(5) It is clinically advisable on occasion to advise a woman to have a child.

(6) The loss of a baby by miscarriage, stillbirth, or neonatal death occasions instinctive deep distress and grief.[15]

In contrast, a quite different interpretative repertoire is applied to persons categorised as 'unmarried woman'.

(1) Pregnancy and childbearing are abnormal and undesirable and conversely the desire to have a baby is aberrant, selfish, and in need of explanation.

(2) Pregnancy and childbearing are problematic, and not to treat them as such indicates that something is wrong.

(3) Illegitimate children should be surrendered for adoption and a mother who wants to keep her child is unrealistic and selfish.

(4) Diagnostic attention and treatment for infertility is not clinically advisable or relevant – unless the woman is about to get married. It is not proper for her to adopt a child.

(5) It would be most inadvisable and inappropriate clinically to advise a single woman to have a child.

(6) The loss of a baby by miscarriage, stillbirth, or neonatal death should not occasion too much grief or distress, and may even produce relief.[16]

These contrasting pregnancy repertoires resemble the repertoires observed in science in that, in both cases, distinct forms of interpretation are available which appear to express opposed interpretative principles, which on occasion generate problems of interpretative inconsistency, but which are on the whole used successfully by participants to construct interpretations of social action which are adequate for practical purposes. These pregnancy repertoires, however, are likely to be more generally available than those we have identified in science. They are, for example, part of the linguistic potential of pregnant women themselves and they are probably available to all linguistically competent members of British society. It is presumably this familiarity with the two basic pregnancy repertoires and with their varying relevance to the categories 'married and unmarried woman' which enables (most of) us to read 8C as a joke about an unmarried pregnant woman. For, although few readers are likely to fail to 'see the joke', no mention is actually made of pregnancy.

The implicit character of the joke strengthens our confidence in MacIntyre's overall analysis, because it shows clearly that the topic of pregnancy is recognisable solely from the use of standardised phrases from the pregnancy repertoires, that is, the phrases 'good news' and 'bad news', along with the redefinition of the patient which is revealed by the replacement of 'Mrs Brown' with 'Miss Brown'. It is precisely our own familiarity with the discursive regularities documented systematically by MacIntyre that enables us to read the joke as a joke about pregnancy. And it is the sudden switch from one interpretative repertoire to another which creates the incongruity essential to humour.

MacIntyre's analysis supports our own in several respects. It shows that interpretative repertoires can be discerned in areas of social life other than science. It also shows that such repertoires can provide the raw material for non-scientific as well as scientific jokes. In addition, it further confirms that analytical conclusions about serious discourse can be checked by an examination of naturally occurring humour and therefore that humour, rather than being a sociological frivolity, is a topic of critical sociological significance.

Humour and the real social world

In one respect, it is clear that humour is a topic which is particularly amenable to analysis in terms of repertoires of discourse. For much humour deals with forms of word-play about an imaginary world involving people and events which do not exist outside the setting of the joke. Sociologists of science, of course, have customarily been concerned with scientists' words only as a source of information about their *real* actions and their *actual* beliefs. Not all humour, however, deals with imaginary events. Sometimes it takes 'the real social world' as its subject. Particular people and their actions can be *made to appear funny*. MacIntyre's study, for example, shows that events very similar to that represented in the pregnancy joke do actually occur. Clearly, an event of this kind can just as easily be portrayed, not as humorous, but as pathetic or distressing. Whether it is laughable or upsetting depends on the telling; and this will vary with the occasion and the participants involved. As we have seen above, humour is created when participants' accounts of action and belief are appropriately organised. Even when actual events are being used as resources for humour, its successful accomplishment depends crucially on the way in which the speaker's version of events is constructed.

Humour, then, is not a characteristic of events in themselves, but is an outcome of the ways in which participants portray and organise their versions of events in the course of social interaction. Participants treat as humorous those interpretative products which have recognisable kinds of structural characteristics; and they produce humour by organising their accounts of action so as to display these characteristics. We have tried to identify above some of the characteristics of scientific humour by examining examples of scientists' humour which have no direct connection with specific events. We suggest, however, that these conclusions are likely to apply equally to humour which is more closely based on actual events. Further study will, of course, be needed to establish this point more firmly.

If this argument is tentatively accepted, we are led to ask whether there is any difference *in principle* between the social production of humour and the social production of, say, consensus, refutation or controversy. In other words, are the phenomena traditionally investigated by sociologists and philosophers of science best conceived, like humour, as aspects or outcomes of the interpretative devices used by scientists to organise their versions of events? Is a controversy analytically different in this respect from a joke? Are the participants' statements used by the analyst to identify and interpret a controversy any less a members' interpretative and context-dependent achievement than is a joke? If the answer is that they are no different in this respect, it appears that we should carry out systematic analysis of traditional topics in the same way that we have begun our analysis here of humour; that is, we should attempt to describe the interpretative procedures used by participants as they construct the discourse through which recognisable social meanings are achieved. This is exactly what we have tried to do in the main body of this book.

Participants themselves, of course, sometimes distinguish between their literal descriptions of social action, which can ostensibly be accepted by the analyst at face value, and the interpretations of action they have devised for specific contexts. For instance, they may compare the versions of action and belief they have produced for the formal literature with their informal, contingent accounts of 'what really happened in the lab'. But such a distinction between literal and constructed versions cannot be adopted by the analyst. For participants treat different versions as literal on different occasions and in different contexts. In other situations, for instance, they will give their formal, empiricist version of events interpretative primacy and treat their everyday, contingent accounts as irrelevant.

There appears, therefore, to be no difference in principle between participants' production of serious discourse and their production of humour. Participants' organisation of humorous discourse is simply one aspect of their ability to construct diverse interpretations of their social world. However, humour is a crucial sociological topic because it involves participants in drawing on their interpretative resources to create, and indeed to celebrate, the kind of discursive variability which, in serious discourse, is largely hidden from the casual observer by the use of reconciliation devices and by the separation of interpretative contexts.

9

•••

Pandora's bequest

In this book we have approached the social world of science as a multiple reality.[1] We have abandoned the traditional sociological goal of producing a single, coherent account of the patterns of action and belief in science. We have sought instead to document some of the methods by means of which scientists construct and reconstruct their actions and beliefs in diverse ways.

At first sight, it may have appeared that, like Pandora, we were heading for chaos. But, as in Pandora's box, Hope still remained; in our case, hope of creating order out of diversity. Although we emphasised that the multiplicity of voices with which scientists and other social actors speak makes traditional sociological objectives unattainable, we held fast to the assumption that interpretative regularities could be discerned behind the babble of tongues, if a suitable analytical approach could be devised. In this book, we have tried to take a few, short steps towards developing such an approach and towards demonstrating what it can tell us about science.

We claim to have shown that scientists use distinctive interpretative forms as they construe their actions and beliefs in different social contexts. We have made an attempt to capture various significant facets of these interpretative forms by devising the concepts of empiricist and contingent repertoires. These concepts have proved to be useful, not only in describing certain recurrent features of scientists' formal and informal discourse, but also in understanding interpretative phenomena which have no obvious connection with our initial observations on versions of action in research papers and interviews. Thus we showed that the two repertoires were used by participants as resources for constructing asymmetrical accounts of error and correct belief. In addition, it became clear that the interpenetration of the two repertoires in interview talk sometimes generated interpretative problems which were resolved by the introduction of the 'truth will out device'. In the first half of the book, therefore, our analysis proved to be fruitful in revealing two basic registers through which scientists are able to create interpretative diversity, in showing how these registers provide the means for constructing major interpretative contexts in science, and in identifying some of the main principles involved in scientists' accounting practices. We were able to show clearly that, although participants' substantive accounts of action and belief are highly diverse, they are constructed out of recurrent interpretative forms and repertoires which can be identified, described and documented by the analyst.

Once we had established these basic conclusions, we moved on to more complex and novel topics. We showed that discourse analysis is not restricted to the realm of small-scale social phenomena. We focused on the supposedly collective phenomenon of cognitive consensus. We argued that it is analytically misleading to treat consensus as a potentially measurable attribute of social collectivities. Sociological analysis along these lines merely serves to reify particular, contextually produced interpretations generated by participants. Examination of participants' interpretative work showed unequivocally that a given collectivity at a given moment can be made to exhibit radically different 'degrees of consensus'. We suggested, therefore, that analytical attention should be directed towards the contextually related methods through which participants construed collective belief as consensual or otherwise. This approach to the study of consensus produced several preliminary findings which could provide the basis for a significantly new interpretative analysis of 'collective belief' in science and of collective phenomena more generally. Although such an approach has not previously been applied to collective phenomena in science we have, of course, been building upon a growing body of interpretative analysis of social aggregates in other areas of social life.[2]

Our next step was to extend the analysis to include types of data which had remained outside the scope of more customary sociological approaches. We did this, in the first instance, by showing that pictorial discourse was open to broadly the same kind of analysis as that which had already been applied to texts and interview transcripts. In particular, it was evident that pictorial versions of scientific knowledge-claims varied in regular ways as they moved between interpretative contexts. Thus, not only does discourse analysis open up for empirical investigation topic areas which had previously been closed, but it shows that such topics can unexpectedly shed new light on longstanding sociological issues. For our analysis showed that examination of scientists' pictures provided an elegant and effective way of dealing with a central problem in the sociology of knowledge; namely, that of clearly demonstrating the contextual variability of scientists' technical representations of the physical world.[3]

This pictorial analysis was linked to our earlier observations on verbal discourse through an examination of scientists' own interpretations of pictures. We found that their interpretations drew on realist and fictionalist conceptions of pictorial representation which closely paralleled the empiricist and contingent repertoires they used to portray action and belief. Their discourse about pictures, however, was overwhelmingly fictionalist in character. The one major exception to this in our data was in their talk about pictures which they said were intended for students and for popular consumption. These pictures, they stressed, had to be constructed in more realistic terms; and our examination of such pictures showed that they frequently were different from the pictures circulated among specialists and that they often included more 'realistic' visual components drawn from the realm of everyday representations of ordinary objects. It also became clear that scientists' use of and movement between fictionalist and realist repertoires in talking about pictures frequently created interpretative problems which were similar to those which appeared in transitions between empiricist and contingent discourse about action and belief.

The most evident of these interpretative problems, namely, 'Trubshaw's dilemma', arose out of scientists' difficulty in reconciling their fictionalist accounts of pictures with their claim that more realistic pictures were suitable for students. We showed that this interpretative problem was not confined to our respondents' reflective talk about pictures, but that the dilemma reappeared in the visual domain itself. This point was strengthened by looking at a form of visual joke in which components 'not to be taken seriously' are represented humorously by means of pictorial resources taken from a quite different area of discourse. Visual jokes of this kind could be seen to resolve Trubshaw's dilemma by being organised to provide a clear guide to the 'degree of realism' to be attributed to the components of the knowledge-claim in question.

The oxidative phosphorylation cartoon led us to take scientists' jokes seriously. We have treated them as a form of discourse in which participants' potential interpretative diversity is clearly revealed. We have employed them, therefore, as a check upon our prior conclusions and, by selecting jokes which have wide currency within the scientific community, we have used them to show that some at least of our findings are relevant to naturally occurring discourse among scientists in general.

We also stressed that the peculiar analytical usefulness of humour is not restricted to science; and we illustrated this with a brief digression to consider MacIntyre's analysis and the pregnancy joke. But it is not just our use of humour which is of general sociological significance. For our basic argument presented in chapter one, that traditional forms of sociological analysis of action are derived in an unexplicated fashion from participants' discourse and that discourse analysis is a necessary prelude to, and perhaps replacement for, the analysis of action and belief, is a completely general argument which applies equally to all areas of sociological inquiry. We hope, therefore, that this book will be read, not simply as an attempt to give further momentum to a new approach under way within the sociology of science, but as a contribution to a wider analytical movement in sociology and in other disciplines concerned with the production and reproduction of social life through discourse.[4]

Notes

1 Scientists' discourse as a topic

1 A systematic exposition based on this literature is provided in Michael Mulkay, *Science and the Sociology of Knowledge*, London: Allen and Unwin, 1979. Some more recent papers can be found in *Knowledge and Controversy: Studies of Modern Natural Science*, a special issue of *Social Studies of Science* edited by H. M. Collins, vol. 11, no. 1, 1981.

2 Virtually all textbooks on social research methods are designed to tell the reader how to obtain the best, single account of the actions which he or she is investigating. This is true even of those texts where great emphasis is placed upon 'going to the people' and letting them speak for themselves. See, for example, R. Bogdan and S. J. Taylor, *Introduction to Qualitative Research Methods*, New York: Wiley, 1975.

3 The nature of the analysts' definitive versions of social action is examined in Michael Mulkay, 'Action and belief or scientific discourse? A possible way of ending intellectual vassalage in social studies of science', *Philosophy of the Social Sciences*, vol. 11, 1981, pp 163–71.

4 John Heritage, 'Aspects of the flexibilities of natural language use: a reply to Phillips', *Sociology*, vol. 12, 1978, pp 79–103.

5 Marlan Blissett, *Politics in Science*, Boston: Little, Brown and Co, 1972, especially pp 138–43.

6 Ibid., p 138.

7 Ibid., p 139.

8 Ibid., p 141.

9 Ibid., p 142.

10 G. Nigel Gilbert and Michael Mulkay, 'In search of the action: some methodological problems of qualitative analysis', in *Accounts and Action*, edited by G. Nigel Gilbert and Peter Abell, Aldershot: Gower, 1983.

11 Michael Mulkay, Jonathan Potter and Steven Yearley, 'Why an analysis of scientific discourse is needed', in *Science Observed: Contemporary Analytical Perspectives*, edited by Karin Knorr-Cetina and Michael Mulkay, London and Beverly Hills: Sage, 1983.

12 M. A. K. Halliday, *Language As Social Semiotic*, London: Edward Arnold, 1978, pp 28–9 and 32.

13 Mulkay, 'Action and belief or scientific discourse?'

14 This kind of problem is explored systematically in J. D. Douglas, *Investigative Social Research*, Beverly Hills and London: Sage, 1976.

15 For other symbolic domains and their relationship to language, see Nelson Goodman, *Languages of Art*, Brighton: Harvester Press, 1981. We will begin to explore the connections between language and pictorial representation in chapters seven and eight.

16 This work is reviewed by Karin Knorr-Cetina, 'The programme of constructivism in science studies: theoretical challenges and empirical results of ethnographies of scientific work', in *Science Observed*, edited by Knorr-Cetina and Mulkay.

17 Bruno Latour and Steve Woolgar, *Laboratory Life: The Social Construction of Scientific Facts*, London and Beverly Hills: Sage, 1979, chapter 3.

18 Norwood Russell Hanson, *Patterns of Discovery*, Cambridge: Cambridge University Press, 1958; G. Nigel Gilbert and Michael Mulkay, 'Contexts of scientific discourse: social accounting in experimental papers', pp 269–94 in *The Social Process of Scientific Investigation*, edited by K. Knorr *et al.*, Dordrecht/Boston: Reidel, 1980. This paper forms the basis for chapter three below.

19 Alfred Schutz, *The Phenomenology of the Social World*, London: Heinemann, 1972.

20 H. M. Collins and T. J. Pinch, *Frames of Meaning*, London: Routledge and Kegan Paul, 1982.

21 Collins is the most enthusiastic advocate of this kind of approach in the sociology of science. See his 'Respondents' talk and participatory research', a paper given at the University of Surrey Accounts of Action Conference, December 1981. For a general discussion of the craft element in social research, see C. Wright Mills, *The Sociological Imagination*, Oxford: Oxford University Press, 1959.

22 G. Nigel Gilbert and Michael Mulkay, 'Warranting scientific belief', *Social Studies of Science*, vol. 12, 1982, pp 383–408; 'Scientists' theory talk', *Canadian Journal of Sociology*, vol. 8, 1983, pp 179–97; and 'Opening Pandora's Box', *Sociology of Arts and Sciences*, forthcoming.

23 For example, James A. Beckford, 'Accounting for conversion', *British Journal of Sociology*, vol. 29, 1978, pp 249–62 and 'Talking of apostasy and "telling" tales' in *Accounts and Action*, edited by G. Nigel Gilbert and Peter Abell, Aldershot: Gower Press, 1983. Jonathan Potter, 'Nothing so practical as a good theory: the problematic application of social psychology', in *Confronting Social Issues: Applications of Social Psychology*, vol. 1, edited by Peter Stringer, London: Academic Press, 1982; Shirley Prendergast and Alan Prout, 'What will I do . . .? Teenage girls and the construction of motherhood', *Sociological Review*, vol. 28, 1980, pp 517–35.

24 Detailed discussion of actual instances of these kinds of problem can be found in Jonathan Potter and Michael Mulkay, 'Scientists' interview talk: interviews as a technique for revealing participants' interpretative practices' in *The Research Interview: Uses and Approaches*, edited by M. Brenner *et al.*, London: Academic Press, 1982.

25 Michael Mulkay and G. Nigel Gilbert, 'What is the ultimate question? Some remarks in defence of the analysis of scientific discourse', *Social Studies of Science*, vol. 12, 1982, pp 309–19.

26 Mulkay, Potter and Yearley, 'Why an analysis of scientific discourse is needed'; Steve Woolgar, 'Interest and explanation in the social study of science', *Social Studies of Science*, vol. II, 1981, pp 365–94.

27 We have included in the following list all the studies known to us which contain some form of sociological analysis of scientific discourse, apart from those cited elsewhere in this chapter. D. C. Anderson, 'Some organisational features in the local production of a plausible text', *Philosophy of the Social Sciences*, vol. 8, 1978, pp 113–35; Charles Bazerman, 'What written knowledge does: three examples of academic discourse', *Philosophy of the Social Sciences*, vol. II, 1981, pp 361–87; Charles Bazerman, 'Forces and choices shaping a scientific paper: Arthur H. Compton, physicist as writer of non-fiction', paper presented at the Sixth Annual Meeting of the Society For Social Studies of Science, Atlanta, November 1981; Augustine Brannigan, *The Social Basis of Scientific Discoveries*, Cambridge: Cambridge University Press, 1981; Michel Callon, J. P. Courtial and W. Turner, 'Co-word analysis: a new method for mapping science and technology', mimeo, Université Louis Pasteur, GERSULP, Strasbourg; G. Nigel Gilbert, 'Referencing as persuasion', *Social Studies of Science*, vol. 7, 1977, pp 113–22; Joseph Gusfield, 'The literary rhetoric of science: comedy and pathos in drinking driver research', *American Sociological Review*, vol. 41, 1976, pp 16–34; Karin Knorr-Cetina, *The Manufacture of Knowledge*, Oxford: Pergamon Press, 1981; Bruno Latour and P. Fabbri, 'La rhetorique du discours scientifique', *Actes de la Recherche en Sciences Sociales*, vol. 13, 1977, pp 81–95; Michael Lynch, *Art and Artifact in Laboratory Science: A Study of Shop Work and Shop Talk in a Research Laboratory*, London: Routledge and Kegan Paul, 1984; K. L. Morrison, 'Some properties of "telling-order designs" in didactic inquiry', *Philosophy of the Social Sciences*, vol. II, 1981, pp 245–62; Michael Mulkay, 'Norms and ideology in science', *Social Science Information*, vol. 15, 1976, pp 637–56; Michael Mulkay, 'Interpretation and the use of rules: the case of the norms of science', pp 111–25 in *Science and Social Structure: A Festschrift for Robert Merton*, edited by Thomas Gieryn, Transactions of the New York Academy of Sciences, series II vol. 39, 1980; Nicholas C. Mullins, 'Paper, forms and groups', paper presented at the Sixth Annual Meeting of the Society for Social Studies of Science, Atlanta, November 1981; Jonathan Potter, P. Stringer and M. Wetherell, *Social Texts and Contexts: Literature and Social Psychology*, London: Routledge and Kegan Paul, 1983; Steve Woolgar, 'Writing an intellectual history of scientific development: the use of discovery accounts', *Social Studies of Science*, vol. 6, 1976, pp 395–422; Steve Woolgar, 'Changing perspectives: a chronicle of research development in the sociology of science', in *Sociology of Science and Research: Papers of the International Sociology of Science Conference Budapest 1977*, Budapest: Akademiai Kiado, 1979; Steve Woolgar, 'Discovery: logic and sequence in a scientific text', pp 239–68 in *The*

Social Process of Scientific Investigation, Dordrecht/Boston: Riedel, 1980; Steven Yearley, 'Textual persuasion: the role of social accounting in the construction of scientific arguments', *Philosophy of the Social Sciences*, vol. 11, 1981, pp 409–35; Steven Yearley, *Contexts of Evaluation: A Sociological Analysis of Scientific Argumentation with reference to the History·of Earth Science*, D. Phil. thesis, University of York, England.

28 Frank Burton and Pat Carlen, *Official Discourse: On Discourse Analysis, Government Publications, Ideology and the State*, London: Routledge and Kegan Paul, 1979.

29 Malcolm Coulthard and Martin Montgomery (eds.), *Studies In Discourse Analysis*, London: Routledge and Kegan Paul, 1981.

30 Peter Roe, *Scientific Text: Selections from the Linguistic Evidence Presented in a Study of Difficulty in Science Text-books*, Discourse Analysis Monographs no. 4, English Language Research, Birmingham University, 1977.

31 Once again, this is no more than a difference of emphasis between ourselves and the sociolinguists. For example, Coulthard, Montgomery and Brazil express a similar approach to the structural concepts of sociology when they state that: 'While these factors are quite obviously defined in sociological terms the investigation remained a linguistic one for, while initially it was essential to use concepts like "status", and talk in terms of social roles like "chairman", the hope was that eventually it would be possible to come full circle and define roles like "chairman" as a set of linguistic options.' Coulthard and Montgomery, *Studies in Discourse Analysis*, p 14.

32 David Brazil, 'The place of intonation in a discourse model', pp 146–57 in *Studies in Discourse Analysis*.

2 A possible history of the field

1 Our strategy for dealing with interview transcripts corresponds roughly with that advocated by J. Lofland, *Analyzing Social Settings*, Belmont: Wadsworth, 1971.

2 A description of co-citation analysis may be found in Henry Small and Belver C. Griffith, 'The structure of scientific literatures, I and II', *Science Studies*, vol. 4, 1974, pp 17–40 and 339–65.

3 Contexts of scientific discourse

1 M. A. K. Halliday, *Language As Social Semiotic*, London: Edward Arnold, 1978, p 189 (emphasis added).

4 Accounting for error

1 David Silverman, 'Interview talk: bringing off a research instrument', *Sociology*, vol. 7, 1973, pp 31–48

2 Crosskey, in passing, seems to be constructing here an incipient account of his own earlier errors in the light of his current view of chemiosmosis. His

explanatory resource is the notion of 'dogmatism', which occurs in many of our examples.

3 Melvin Pollner, 'Mundane reasoning', *Philosophy of the Social Sciences*, vol. 4, 1974, pp 35–54.

4 Ibid., p 39.

5 Ibid., p 48 (italics in the original).

5 The truth will out

1 H. Sacks, E. Schegloff and G. Jefferson, 'A simplest systematics for the organisation of turn-taking for conversation', *Language*, 1974, pp 696–735.

2 Bruno Latour and Steve Woolgar, *Laboratory Life*, Beverly Hills and London: Sage, 1979.

6 Constructing and deconstructing consensus

1 This issue is discussed in more general terms in *Advances in Social Theory and Methodology: Toward an Integration of Micro and Macrosociologies*, edited by K. Knorr-Cetina and A. Cicourel, London: Routledge, 1982.

2 In order to safeguard participants' anonymity as far as possible, we will not provide references for the published sources used in this chapter.

3 John Ziman, *Public Knowledge: The Social Dimension of Science*, Cambridge: Cambridge University Press, 1968, p 9.

4 Karin D. Knorr, 'The nature of scientific consensus and the case of the social sciences', pp 227–56 in *Determinants and Controls of Scientific Development*, edited by K. Knorr et al., Dordrecht: Reidel, 1975, pp 252–3 (italics in the original).

5 Ibid., p 242.

6 For a detailed study of how scientists reach agreement informally, see Michael Lynch, *Art and Artifact in Laboratory Science: A Study of Shop Work and Shop Talk in a Research Laboratory*, London: Routledge and Kegan Paul, 1984.

7 For a discussion of some of scientists' folk theories, see Augustine Brannigan, *The Social Basis of Scientific Discoveries*, Cambridge: Cambridge University Press, 1981.

8 These recurrent features are not, of course, observable apart from the interpretative work carried out by ourselves as analysts or by some other hearers or readers. However, we have shown that the interpretative procedures which we have identified are significant elements in participants' discourse by documenting how they become the focus of participants' own efforts at interpretative deconstruction.

9 See Michael Mulkay, 'Consensus in science', *Social Science Information*, vol. 17, 1978, pp 107–22.

7 Working conceptual hallucinations

1 Robert A. Day, *How to Write and Publish a Scientific Paper*. Philadelphia: ISI Press, 1979.

2 J. B. Finean, R. Coleman and R. H. Michell, *Membranes and their Cellular Functions*, Oxford: Blackwell, 1978.

3 Martin J. S. Rudwick, 'The emergence of a visual language for geological science 1760–1840', *History of Science*, vol. 14, 1976, pp 149–95.

4 Jerome R. Ravetz, *Scientific Knowledge and its Social Problems*, Oxford: Clarendon Press, 1971.

5 This is, of course, a pseudonym. For the same reason as before we provide no specific reference.

6 Finean *et al.*, *Membranes and their cellular functions*, p 90.

7 Ibid., p 73.

8 For other examples of visual jokes in a textbook, see David G. Nicholls, *Bioenergetics: An Introduction to the Chemiosmotic Theory*, London and New York: Academic Press, 1982. The use of 'unrealistic' components in textbook cartoons is not rigidly restricted to the representation of phenomena which are defined as 'not yet understood'. Such components can also be used to represent phenomena which are treated as 'not directly relevant to' the topic in question; even though these phenomena are taken to be well understood and as amenable, in principle, to much more 'realistic' representation in appropriate circumstances.

8 Joking apart

1 See for example the articles in *Impact of Science on Society*, vol. 19, no. 3, 1969, which is devoted entirely to scientists' humour.

2 *Hopkins and Biochemistry*, Cambridge: Heffer and Sons, 1949.

3 E. Garfield, 'Humour in scientific journals, and journals of scientific humour', *Essays of an Information Scientist*, vol. 2, Philadelphia: ISI Press, 1977, pp 664–71.

4 I. J. Good *et al.* (eds.), *The Scientist Speculates*, New York: Basic Books, 1962; *A Random Walk in Science*, an anthology of scientists' humour compiled by R. L. Weber, London and Bristol: the Institute of Physics and New York: Crane, Russak, 1973.

5 Sidney Harris, *What's So Funny About Science?*, Los Altos, California: William Kaufmann, 1980.

6 Harold Baum, *The Biochemists' Songbook*, Oxford and New York: Pergamon Press, 1982.

7 Two of the very few analysts to treat scientific humour as a fruitful topic are Martin J. S. Rudwick, 'Caricature as a source for the history of science: De La Beche's anti-Lyellian sketches of 1831', *Isis*, vol. 66, 1975, pp 534–60 and G. D. L. Travis, 'On the construction of creativity: the "memory transfer" phenomenon and the importance of being earnest', pp 165–93 in *The Social Process of Scientific Investigation*, edited by K. Knorr *et al.*, Dordrecht and Boston: Reidel, 1980.

8 John Allen Paulos, *Mathematics and Humour*, Chicago and London: The University of Chicago Press, 1980, p 9.

9 An attempt to develop a sociological analysis of irony can be found in Edmund Wright, 'Sociology and the irony model,' *Sociology*, vol. 12, 1978, pp 523–43.

10 For a detailed examination of the place of laughter in certain passages of discourse, see Gail Jefferson, 'A technique for inviting laughter and its subsequent acceptance declination', pp 79–95 in *Everyday Language: Studies in Ethnomethodology*, edited by G. Psathas, New York: Irvington, 1979.

11 Good, *The Scientist Speculates*, pp 52–3; also in Weber, *A Random Walk in Science*, pp 120–1.

12 Weber, *A Random Walk in Science*, p 140.

13 Ibid., pp 167–8.

14 Sally MacIntyre, ' "Who wants babies?" The social construction of "instincts" ', pp 150–73 in *Sexual Divisions and Society*, edited by Diana Barker and Sheila Allen, London: Tavistock, 1976.

15 Ibid., p 159.

16 Ibid., pp 159–60.

9 Pandora's bequest

1 For analyses of multiple realities, see Erving Goffman, *Frame Analysis: an Essay on the Organisation of Experience*, New York: Harper and Row, 1974; Melvin Pollner, 'The very coinage of your brain: the anatomy of reality disjunctures', *Philosophy of the Social Sciences*, vol. 5, 1975, pp 411–30; Dorothy Smith, 'K is mentally ill: the anatomy of a factual account', *Sociology*, vol. 12, 1978, 23–53; E. C. Cuff, *Some Issues in Studying the Problem of Versions in Everyday Situations*, Occasional paper no. 3, Department of Sociology, University of Manchester, 1980; Alfred Schutz, *The Phenomenology of the Social World*, London: Heinemann, 1972.

2 For example, Jack D. Douglas, *The Social Meanings of Suicide*, Princeton: Princeton University Presss, 1967; *Understanding Everyday Life*, edited by Jack D. Douglas, London: Routledge and Kegan Paul, 1971. In so far as scientific discovery is a collective phenomenon, two previous contributions which point towards our form of analysis of collective phenomena are Steve Woolgar, 'Writing an intellectual history of scientific development: the use of discovery accounts', *Social Studies of Science*, vol. 6, 1976, pp 395–422 and Augustine Brannigan, *The Social Basis of Scientific Discoveries*, Cambridge: Cambridge University Press, 1981.

3 For this kind of reformulation of the central concern of the sociology of knowledge, see Michael Mulkay, *Science and the Sociology of Knowledge*, London: Allen and Unwin, 1979, p 93.

4 For examples of related work on discourse in disciplines other than sociology, see the references to chapter one. See also S. R. Horton, *Interpreting Interpreting: Interpreting Dickens' 'Dombey'*, Baltimore: John Hopkins University Press, 1979; J. V. Harari (ed.), *Textual Strategies: Essays in Post-Structuralist Criticism*, London: Methuen, 1979; R. Fowler, *Literature as Social Discourse: The Practice of Linguistic Criticism*, London: Batsford Academic, 1981; J. Culler, *The Pursuit of Signs: Semiotics, Literature and Deconstruction*, London: Routledge, 1981; S. Hall, D. Hobson, A. Lowe and D.

Willis, *Culture, Media, Language,* London: Hutchinson, 1980: R. Fowler, B. Hodge, G. Kress and G. Trew, *Language and Control,* London: Routledge, 1979; J. Potter, P. Stringer and M. Wetherell, *Social Texts and Contexts: Literature and Social Psychology,* London: Routledge, 1983.

Index

DATE DUE

DEMCO 38-296

LaVergne, TN USA
15 December 2009
167006LV00004B/8/P

9 780521 274302